McGRAW-HILL SERIES IN AVIATION

David B. Thurston Consulting Editor

DESIGN FOR SAFETY

DAVID B. THURSTON

Consulting Aeronautical Engineer
President, Thurston Aeromarine Corporation

McGRAW-HILL BOOK COMPANY

New York   St. Louis   San Francisco   Auckland   Bogotá   Hamburg   Johannesburg
London   Madrid   Mexico   Montreal   New Delhi   Panama
Paris   São Paulo   Singapore   Sydney   Tokyo   Toronto

This book is dedicated to members of the aviation community who have contributed to accident statistics and records. The misfortunes of this unwilling group have pointed the way toward design and operational improvements benefiting all who fly now and in the years ahead.

Library of Congress Cataloging in Publication Data

Thurston, David B
Design for safety.

(McGraw-Hill series in aviation)
Includes index.
1. Aeronautics—Safety measures. 2. Airplanes—
Design and construction.   I. Title.   II. Series.
TL553.5.T48   629.134'1   79-25241
ISBN 0-07-064554-X

1234567890   VHVH   89876543210

The editors for this book were Jeremy Robinson and Margaret Lamb,
the designer was Elliot Epstein, and the production supervisor
was Teresa F. Leaden. It was set in Optima
by Progressive Typographers.

Printed and bound by Von Hoffman Press, Inc.

Frontispiece: Amphibious floats increase utility and safety by permitting operation from water as well as land.
                                        EDO-AIRE Seaplane Photo

CONTENTS

automatic alternate air door • fuel selector operation • fuel system design • range requirements • detail features • superchargers • propeller and fan design • engine safety instruments.

for small aircraft • runway numbering • improving unicom and control communications • simpler holding patterns • enroute separation • HUD equipment • proximity indicators • weather reporting.

PREFACE

It is virtually impossible to pick up any aviation or weekly newsmagazine without running across some aircraft accident report or related statistic. Not only is much of this information frequently confusing, but any possible type of failure seems to be presented from time to time as the number one cause of general aviation accidents. As a result, an objective study of flight safety was undertaken to learn what major factors really cause trouble—and so determine what can be done to improve our accident record.

During preparation of my earlier book, *Design for Flying,* it became increasingly evident that certain gaps in applied technology along with current design practice and flight procedures contribute to aircraft accidents, particularly those normally attributed to "pilot error."

It follows that if and when these oversights are corrected we will realize increased general aviation safety, accompanied by reduced insurance and aircraft costs inviting greater acceptance of the personal airplane as a utility vehicle. And that is what this book is all about—safer, and cheaper, private flying.

We will first review the principal causes of general aviation accidents in a constructive manner, including a brief analysis of why certain types of accidents are really not so much the result of pilot error as they are the fault of poor design, poor flight instruction technique, inadequate air traffic control and communication procedures, and questionable airport location.

Fortunately, all these factors are subject to some degree of correction and improvement—provided influential members of the aviation community will listen long enough to realize the benefits we all enjoy from a better safety record. To that end, this book includes steps that can be taken to effect needed changes, with all recommendations lying within the current state-of-the-art for modern aircraft design and operation.

The design review indicates that although improved flight safety is easily within our grasp and can command a slightly higher initial cost, safer aircraft will not be accepted unless they are competitive with existing types in performance, handling characteristics, load capacity, range, and construction.

However, as an incentive, the lower accident rate and increased sales demand generated by safe aircraft designs developed in this book should actually lower purchase and operating costs in the years ahead.

Our flight analysis shows where the most frequent types of pilot error accidents occur along the flight profile from takeoff to landing; where we should design some potential accident factors out of existing aircraft and operational procedures; and the impact of an improved safety record upon liability and hull insurance.

To assist in reaching these conclusions, the National Transportation Safety Board supplied basic accident data. Particular appreciation is due Edward E. Slattery, Jr., director of the Office of Public Affairs in Washington, D.C., for statistical, definition, and background material.

Additional general aviation accident information was provided by Capt. Homer Mouden, technical director of Flight Safety Foundation, Inc., Arlington, Virginia, who placed special emphasis upon the effect of density altitude in causing accidents.

The Aircraft Owners and Pilots Association (AOPA) Air Safety Foundation also participated through background data and statistical material forwarded by Vice President Charles Spence.

As before, my wife Evelyn acted as translator and typist, which, along with her editorial comments offered from time to time, helped speed this book toward completion.

Many thanks are also due our son, Roy, whose suggestions made the statistical background more readable and interesting.

Last but by no means overlooked is our collective debt to those who have unwillingly and frequently unknowingly provided accident statistics. Losses experienced over the years by these pilots and their passengers will become aviation's gain if the resulting safety features recommended in this book are used to reduce aviation accidents, thus assuring broader acceptance of the personal airplane as a safe and useful vehicle in the years ahead.

David B. Thurston

PART ONE
THE BASIS FOR CHANGE

INTRODUCTION

Since earliest history, tales of heroic voyages have been filled with interesting problems and dire disasters caused by weather at its worst. And wherever we travel on the sea or through the air, unexpected or unavoidable bad weather remains today's number one villain.

There can be no argument that most accidents are caused by pilot judgment and operational practices, or that weather will remain an incurable cause of a great number of flying accidents. In a constant attempt to reduce these statistics, flight procedures and precautionary measures intended to improve pilot capability have been repeatedly published over the past 65 years. But regardless of this effort, pilot error accidents continue at an almost constant rate, while there is apparently some distance to go before we harness the weather.

As an alternative, I suggest we focus our interest upon a review of general aviation design and operational procedures to determine whether many so-called or recorded pilot error accidents would occur if basic features causing these accidents were eliminated.

The potential for this approach to improve flight safety was realized after a casual survey of accident data and related statistics. During this rather boring and thankless effort, it became quite obvious that some items were causing about the same percentage of trouble every year. Not much can be done about the weather, but according to statistical data our aircraft and flight procedures are consistently poor. What can we do to improve this situation?

Rather than attempting to modify human nature by endless discussion of the dos and the don'ts of flying, it seems time we gave some thought to removing as many as possible of the operational procedures and design features that might contribute to pilot error. Simply stated: *A control or procedure that does not exist cannot be improperly managed,* and this is particularly important during periods of extreme or unexpected emotional stress.

The engine stopping, ice forming rapidly, running into severe clear-air turbulence (CAT) or an imbedded thunderstorm, or having a complete electrical system failure in instrument conditions will cause emotional stress regardless of anyone's flying time or years of piloting experience. Proper operation of every one of many procedures and gadgets under such extreme situations may be beyond human capacity—and so an accident results.

If we can reduce demands upon the pilot during normal as well as emergency conditions by improving aircraft design and required flight procedures, the opportunities for pilot error will be correspondingly reduced along with the accident rate.

In view of this, Part Two, Safety by Design, focuses upon pilot error accidents that could be eliminated by different design procedures, including a basically new aircraft design philosophy. Development specifications and justification for producing this new type of airplane are covered in Chapters 3 through 7.

Since flight qualities and design criteria are directly related, and therefore virtually inseparable, Part Three, Flight Safety, covers recommendations for preventing pilot error–provoking situations directly related to flight procedures, airports, and airways.

Taken together, the combination of possible design and flight improvements that can be realized within current state-of-the-art techniques is shown capable of materially reducing the accident rate. In addition to saving lives, a reduced accident rate would make aircraft use and ownership more attractive and less costly for everyone.

This goal is worth serious consideration and implementation, but it can be realized only if those of us interested in aviation argue for the necessary changes. One reason this book was written was the hope that many readers will work to obtain recommended improvements relating to individual areas of interest, thereby personally contributing to the next generation of better and safer aircraft, airways, airports, and flight procedures.

In preparing this book, every effort was made to use the full spelled-out form for any organizational name, procedure, or technical phrase before using an abbreviated form. For your convenience, all such abbreviations and acronyms have been gathered into a glossary preceding the index, permitting ready reference if memory becomes rusty or the acronyms too similar as you progress through the text.

1

REVIEWING ACCIDENT DATA

After an accident occurs, the accounting begins. Flight conditions, pilot experience, and aircraft data collected during each accident investigation are analyzed by National Transportation Safety Board (NTSB) personnel with assistance from the Federal Aviation Administration (FAA) as required. The NTSB issues detailed reports of its findings for each accident and publishes the *Annual Review of Aircraft Accident Data* summarizing the year's results. A study of this information will then define the need for specific design and operational safety recommendations presented in Chapter 2. But first we should briefly consider accident definitions and terminology along with the different categories of accident data.

To provide additional background for interested readers, a review of safety agencies and accident investigation procedures has been included in the Appendix following the final chapter.

ACCIDENT DEFINITIONS AND CATEGORIES

If you have ever read a copy of the NTSB's *Annual Review of Aircraft Accident Data*, certain terms may have seemed confusing, frequently leading to unanswered questions. In view of this, the following definitions and discussion are offered to help clarify accident terminology; some were obtained in response to correspondence with the NTSB, while others are standard in the aircraft industry. All the terms appear regularly in accident reports and will be used throughout this book.

Aircraft accident means an occurrence associated with the operation of an aircraft in which any person suffers death or serious injury as a result of being in or upon the aircraft, or in which the aircraft receives substantial damage. (Any reader wishing to examine this definition in greater detail will find a lengthy legal explanation in Ref. 1.2, although the above shorter form satisfies our requirements.)

Type of accident indicates *what happened* under the immediate circumstances involved in the accident. One incident resulting in substantial damage can involve more than one type of accident. For example, a stall resulting in collision with the ground involves two types of accidents. The stall would be listed as the first type of accident which then produced the second—collision with the ground. Both events would be included in NTSB statistics, which explains why the total number of detail causes exceeds the actual number of accidents reported for any given period. As shown later in Figure 1-1 for 1970, although 4290 small fixed-wing aircraft accidents were recorded for that year, there were 10,373 different cause-and-factor items associated with those accidents.

Airframe refers to the basic load-carrying structure of an airplane; this includes the wings, tail surfaces, fuselage or hull, cowling, engine mount, control system, and landing gear. Each of these components represents a portion of the airframe.

Airframe failures in flight are occurrences resulting from failure of any part of the airframe during flight, regardless of the cause. Most accidents of this type occur during penetration of weather and result from maneuvers exceeding aircraft structural design load specifications. The cause most frequently cited is "continued VFR flight into adverse weather conditions," a pilot error situation frequently resulting in spatial disorientation due to pilot inexperience and ending with a spiral or spinning descent.

Mismanagement of fuel (a pilot error item) can cause fuel starvation as well as improper weight distribution.

Fuel starvation occurs when ample fuel is aboard the aircraft but for some reason the flow of fuel to the engine is interrupted, reduced, or completely stopped.

Nose-over/-down accidents might sound as though a stall or spin were involved; however, these are actually ground operation mishaps. *Nose down* describes an airplane accidentally turning up on its nose while on the ground or water during surface maneuvers. *Nose over* refers to an aircraft accidentally turning up on its nose and continuing forward rotation until it ends up on its back on the runway, off-field emergency landing area, or water surface.

Stall or spin accidents occur while aircraft are in airborne maneuvers. The *stall* is a condition in which either the airflow separates from the airfoil surface or the airflow around the airfoil becomes turbulent, causing loss of wing lift. The result is a loss of altitude until recovery can be effected. The *spin* is a maneuver of a stalled aircraft, either controlled or uncontrolled, in which the airplane descends in a helical path at a wing angle of attack greater than the angle of maximum lift—meaning the wing is stalled and not producing lift.

Spirals differ from spins in that the wing angle of attack remains within the normal range of flight angles—so the wing continues to provide lift without stalling.

Cause/factor items and types of accident, difficult areas to understand, need an example. As would be expected, when accidents involve a student on solo flight, the student is listed as the pilot in command. When both instructor and student are aboard, the instructor is reported as pilot in command. During dual-instruction operation accidents, either the student or the instructor or both can be listed as a cause/factor in the accident. This means that if the student overshoots and the instructor does not react in time to prevent a collision with trees, the *first type of accident* will be listed as an overshoot, with the *second type of accident* being in the "collision with" category under the item "trees" to explain what happened. In addition, both the instructor and student will be listed as cause/factor items—they were the *factors* that *caused* the accident.

At first the various categories of types of accident plus the many cause/factor items listed in the published statistics seem confusing, although review of the preceding example will help clarify these terms. The basic *accident type categories* include: surface accidents, collision between aircraft, collision with the surface from flight, collision with numerous obstacles other than the earth, stall, fire or explosion, airframe failure, miscellaneous, and undetermined cause. Most categories have a number of subdivisions resulting in a total of about 60 different listed *types of accident*.

A further breakdown of accident data is provided by the *cause/factor summaries* of numerous items under the general headings of pilot, copilot, dual student, check pilot, personnel (instructor, maintenance, weather, traffic control, etc.), airframe, powerplant, systems (electrical, hydraulic, etc.), airports/airways/facilities, weather, terrain, and miscellaneous. All these listings include some 350 different cause/factor items, making for comprehensive coverage of every accident reviewed by the NTSB. Incidentally, items carried in the cause/factor listings under pilot, copilot, student, and check pilot are the source references for pilot error statistics we hear so much about.

Since wading through all this detail becomes a monumental task, the most frequent accidents plus those over which we can exercise some control have been extracted for a number of different years and presented in the following section as simple tables for comparison. A review of this statistical data provides some interesting facts, particularly in regard to: instrument flight rules (IFR) operation, powerplant problems, airport and airways troubles, and accidents due to lack of airspeed. Weather accidents have been included for comparison purposes, although as previously noted we can do little to improve upon this naturally uncontrollable factor.

STATISTICAL DATA A statistical analysis of general aviation accidents was published annually by the Civil Aeronautics Board (CAB) through 1965. After that date, the NTSB assumed responsibility for accumulating data, classifying accidents, and publishing the *Annual Review of Aircraft Accident Data*. Each annual report is released in final form about 1 year after the reporting period, with interim statistics issued during the year being covered.

Calendar year	Approx. number of active general aviation aircraft	Total of small fixed-wing aircraft accidents reported*	Total active aircraft accidents, %	Total of all cause/factor items*	
1948	77,100	7850	10.2	Not listed	
1970	133,000	4290	3.2	10,373	
1971	136,000	4243	3.1	10,122	
1975	167,000	3738	2.7	10,478	
1976	176,000	3770	2.1	10,293	
1977	185,200	3842	2.4	9,467	

FIGURE 1-1
Total of direct causes and related factors (causes/factors) associated with selected types of accidents. Data for small fixed-wing aircraft during calendar years indicated.
Data based on Ref. 1.1(a)–(f)

* 1948 values represent all reported general aviation accidents without any cause-and-factor items (which were not prepared at that time).

† Percentages are based upon the total of all cause-and-factor items except for 1948 when the total of all accidents was used.

The seemingly dramatic drop in the total number of aircraft accidents as of 1968 was really due to redefinition of the term *substantial damage* as applied to aircraft with a takeoff weight of 12,500 lb or less—designated as *small aircraft* for certification and regulation purposes. While $300 or more damage had been the cost basis for reporting aircraft accidents through 1967, this requirement was replaced by amended regulations defining *substantial damage* as "damage or structural failure which adversely affects aircraft strength, performance, or flight characteristics, and which would normally require major repairs or replacement of the affected component." Upon reflection, current repair costs were probably growing too fast to permit meaningful comparison with early records anyway.

Although the CAB listed 7850 general aviation accidents for 1948, because of the revised reporting procedure the NTSB recorded only 3842 small fixed-wing aircraft accidents for 1977 when there were considerably more aircraft fluttering around the sky. This change makes direct comparison of old and new statistics rather difficult unless the data for each item is listed as a percentage of all accidents reported for the period. This approach has been followed for a random selection of years in preparing Figure 1-1; which lists the total reported causes of certain accidents to small fixed-wing aircraft (the only type considered in this book, including both single- and twin-engine models under 12,500 lb).

The pilot error category seems to bear the burden of greatest number of acci-

Pilot error†		Weather		Powerplant and related systems (not due to pilot error)		Airport and airways facilities (not due to pilot error)		Total listed in table, %†
Num-ber	Per-cent	Num-ber	Per-cent	Num-ber	Per-cent	Num-ber	Per-cent	
5855	74.59	468	5.96	652	8.31	304	3.87	92.73
5862	56.51	1296	12.49	531	5.12	492	4.74	78.86
5770	57.00	1284	12.68	535	5.28	436	4.31	79.27
5954	56.82	1234	11.78	560	5.34	459	4.38	78.32
6043	58.71	1132	11.00	558	5.42	418	4.06	79.19
5495	58.04	1174	12.40	564	5.96	368	3.89	80.29

dents, with recent pilot error items totaling about 5 times the number charged to weather. In view of this, the totals for pilot error accidents have been included in Figure 1-1 along with other major types of accidents. Certain kinds of pilot error over which we have some design or procedural control have been expanded in Figure 1-2 for more detailed consideration. Note that the 1948 data for both figures represent actual accident totals, while the more recent reports carry a listing of all causes as well.

Most accidents involve overlapping items such as: improper operation of powerplant (1), which caused the powerplant to stop running (2), which caused the airplane to glide to a crash landing (3), which damaged the airframe (4), and which injured the pilot (5) and one passenger (6). Thus, this one accident event would be listed in at least six different accident item records plus the record of any other cause/factor, such as fuel mismanagement, which might be determined during the investigation. By this reckoning, it is easy to see why the total number of accident causes and factors exceeds the total number of actual accidents investigated during the reporting period.

In the above example, the types of accident would have been listed as follows: (1) engine failure resulting in (2) a controlled collision with the ground. All subsequent items concerning why the accident happened are tabulated as associated causes and factors of the accident. Of course, if damage were the only criterion, this accident occurred when the airplane hit the ground; but when detailed in NTSB reports, two types of accidents would be listed for this

Calendar year	Total of all cause/ factor items*	Failed to obtain or maintain flying speed†		Failed to see and avoid other aircraft, objects, and obstructions		Mismanage- ment of fuel		Improper operation of powerplant and related systems	
		Num- ber	Per- cent	Num- ber	Per- cent	Num- ber	Per- cent	Num- ber	Per- cent
1948	7,850	569	7.25	433	5.51	233	2.97	328	4.18
1970	10,373	532	5.13	196	1.89	227	2.19	115	1.11
1971	10,122	527	5.21	203	2.00	281	2.78	144	1.42
1975	10,478	495	4.72	258	2.46	255	2.43	108	1.03
1976	10,293	513	4.98	274	2.66	244	2.37	120	1.16
1977	9,467	547	5.78	205	2.16	280	2.95	111	1.17

FIGURE 1-2
Direct causes and related factors (cause/factors) associated with selected pilot error accidents. Data for small fixed-wing aircraft during calendar years indicated.
Data based on Ref. 1.1(a)–(f).

* 1948 values represent all reported general aviation accidents without any cause-and-factor items (which were not prepared at that time).

† Percentages are based upon the total of all cause-and-factor items except for 1948 when the total of all accidents was used.

event: first, engine failure, and second, a controlled collision with the ground.

Getting back to Figure 1-1, we see that weather accidents were not as frequent in 1948 as they have been in recent years, possibly because in 1948 fewer pilots were willing to push their luck in marginal conditions or because the active pilot population was better trained and more skilled right after World War II. But note how weather has remained of major importance even though our aircraft, instruments, communications, and navigational systems have become increasingly sophisticated.

Figure 1-2 clearly shows that the percentage of all accident factors due to continuing under visual flight rules (VFR) into IFR weather is a small part of the total for any one year, although a lethal one according to the fatality records. The same applies to the next two tabulations covering "Improper IFR operation" and "Initiated flight in adverse weather conditions." Despite all we hear about weather accidents caused by pilot error, they represent less than 3 percent of the total for any annual period examined—well below the number attributed to weather alone (without any pilot assistance).

Moving on, we find that when pilot-induced powerplant failures of Figure 1-2 are combined with troubles starting inside the cowling listed in Figure 1-1, all powerplant problems cause about 8 to 9 percent of the total accident items for any year. This interesting rate is exceeded in recent years only by weather, indicating a basic design area requiring further engineering development to reduce pilot workload while providing more reliable powerplant operation.

Improper operation of landing gear retraction systems		Flying VFR into IFR conditions		Improper IFR operation		Initiated flight in adverse weather conditions	
Num-ber	Per-cent	Num-ber	Per-cent	Num-ber	Per-cent	Num-ber	Per-cent
88	1.12	169	2.15	1	0.01	Not listed	
143	1.38	193	1.86	36	0.35	33	0.32
146	1.44	212	2.09	26	0.26	30	0.30
67	0.64	186	1.78	38	0.35	31	0.30
89	0.86	170	1.65	24	0.23	45	0.44
88	0.93	161	1.70	31	0.33	56	0.59

Airport and airway facility accidents of Figure 1-1 also represent a consist-ently steady and considerably greater total than the pilot-induced weather fly-ing troubles shown in Figure 1-2, giving notice that some improvements are overdue in our navigation and airport site selection procedures. So we should look into these segments of flight operation as well.

Failure to obtain or maintain flying speed remains curiously high, as shown in Figure 1-2, even though most aircraft are equipped with stall warning indica-tors. But these are not much help on takeoff, particularly when aircraft are over-loaded or subject to density altitude conditions. Apparently some development is needed to indicate a safe rotation speed for lift-off—and that could be so simple as an adequate takeoff speed vs. weight chart or a mass-flow-speed indi-cator for small aircraft. Something else we shall study later on.

Interestingly enough, Figures 1-2 and 1-3 provide historical proof that pilot error can be decreased by proper design. This is most dramatically shown by the reduction in obstruction and nose-over accidents between 1948 and 1970, the period when tricycle landing gear became firmly established. Conven-tional, or taildragger, gear common in 1948 provided notoriously poor forward visibility when taxiing, even though pilots were supposed to proceed over the ground via a series of slow and shallow S turns. So when tricycle gear offered much safer ground handling, superior crosswind capability, and better forward visibility while taxiing, the flying public accepted a design improvement which immediately reduced collision and nose-over accidents.

Calendar year	Total of small fixed-wing aircraft accidents reported	Powerplant and related systems failure, %	Collision with ground, water, fences, wires, trees, buildings, etc.; controlled and uncontrolled, %	Ground or water loop, %	Stall, spin, spiral, and mush, %	
1948	7850	8.31	33.95	5.44	15.08	
1970	4290	21.98	21.26	14.99	11.12	
1971	4243	22.77	22.11	14.54	10.37	
1975	3738	25.39	24.48	13.56	10.94	
1976	3770	24.26	23.11	13.66	10.87	
1977	3842	23.72	25.13	12.40	11.95	

FIGURE 1-3
Major first types of accidents as a percentage of total accidents. Data for small fixed-wing aircraft (including weather and pilot error accidents).
Data based on Ref. 1.1(a)–(f)

In similar fashion, the introduction of retractable landing gear that automatically lowers at approach airspeed had a major effect upon reducing the frequency of gear-up landings—a pilot error of king-size proportions usually accompanied by endless irritating jests from other pilots (who have yet to make this mistake). As shown by Figure 1-2, pilot error vs. retractable gear accidents (gear up on land or down on water) increased from 1948 to 1971 as did the number of retractable gear aircraft, but then decreased as automatic retraction systems were introduced by Piper on the Arrow and Bellanca with the Viking. The rate may be rising again because of an increase in the number of amphibians, which of course provide repeated opportunity to land on water with the gear down and on land with gear up.

When all pilot error, weather, and other accidents are combined to report the actual number of aircraft accidents by *first type of accident* (what initially happened), we find from Figure 1-3 that the distribution has remained virtually unchanged from 1970 through 1977. Fortunately, the actual accident trend is slowly decreasing even though the domestic small aircraft population is increasing at a net of about 9500 aircraft per year. This obviously represents a most encouraging gain in safety.

The differences in rate between the various types of accidents between 1948 and 1970 and subsequent years are about as would be expected from aircraft design improvements during that period. For instance: the large decrease in nose-over/-down accidents follows introduction of tricycle gear nosewheels as

Hard landing, %	Overshoot, %	Nose-down and nose-over (takeoff or landing), %	Undershoot, %	Total listed, %
12.92	4.67	13.95	2.75	97.07
7.60	4.52	3.45	3.73	88.65
7.80	5.23	3.61	3.25	89.18
6.74	5.35	3.10	2.27	91.83
6.39	5.35	3.67	2.69	90.00
5.34	3.91	3.61	3.58	89.64

previously noted; reductions in surface collisions reflect improved visibility over the nose; fewer hard landing accidents are due to a combination of stronger gear and improved flight handling characteristics; while increases in powerplant problems result from more complex fuel, propeller, and turbo-supercharged systems. However, the increases in ground or water loops are not readily explained unless they indicate differences in reporting procedure, recent growth of seaplane operation, or superior pilot technique when handling conventional landing gear of the 1940s. Also, the grassy airfields of that period tended to be more forgiving and usually permitted operation directly into the wind, thereby reducing crosswind accidents. All these factors are probably responsible for the apparent great leap backward in ground loop accidents between 1948 and 1970—but this still remains an unexpected increase since most aircraft now have tricycle gear which tends to improve ground handling and safety.

Who gets into trouble most often and at what time of day is shown by Figure 1-4. Contrary to popular belief, commercial pilots are no less prone to accidents than private pilots. In fact, their accident percentage exceeds their share of active commercial pilot certificates as shown by Figure 1-4(a), while the private pilot accident rate is lower than the percentage of private pilots. Possibly the relatively high level of commercial pilot accidents results from increased exposure when instructing, dusting, and performing other types of flying that must be self-supporting or that are subject to outside command—such as exec-

Pilot rating	Percent of accidents	Percent of total pilots
Student	11.44	25.15
Private	40.92	41.52
Commercial	29.05	25.23
Airline transport (ATR)	3.12	6.05
Commercial/Flight Instructor	11.69	
ATR/Flight Instructor	2.31	
No certificate	0.95	
Unknown	0.52	
Total	100.00	97.95

(a) Accidents reported by pilot certificate rating.

Time of day and light conditions	Percent of total
Daylight	85.98
Night/dark	7.02
Dusk/twilight	4.05
Night/bright	1.60
Dawn	0.95
Unknown	0.40
Total	100.00

(b) Accidents reported by light conditions.

FIGURE 1-4
Accident frequency related to pilot training and light conditions.
Data based on Ref. 1.1(e) for 1976

utive and small feeder-line operations. It is also quite possible that low-time commercial pilots should not perform these duties; that is another subject for subsequent study.

Most general aviation accidents occur during daytime according to the data gathered in Figure 1-4(*b*). This is not particularly surprising since most small aircraft are flown during daylight hours for a number of reasons including the timing of business appointments, the safety advantage of being able to see what's going on outside the cabin, and natural concern for flying at night with only one engine.

Summing up Figure 1-4 data, we see that about 40 percent of all general aviation accidents are caused by private pilots flying in broad daylight; this is the largest single accident group. As a reassuring note, student pilots have a safety record which is slightly superior to the airline transport rated (ATR) group based upon the percentage of total accidents vs. the percentage of pilots holding student and ATR ratings. So it is apparently relatively safe to learn to fly; trouble develops when inexperienced pilots begin cross-country travel (encountering bad weather, etc.) and when aircraft more complex than basic trainers enter the picture with increased demands upon pilot capability.

THE HOMEBUILT RECORD

Activities of people building their own airplanes (which are commonly referred to as *homebuilts*) focus upon the Experimental Aircraft Association (EAA). Most of these aircraft are constructed as direct copies or personal modifications of existing aircraft for which plans may be purchased, although every year a few are completely new types designed by their builders. A typical aerobatic biplane suitable for homebuilding is shown in Figure 1-5.

Members of the EAA who have been closely associated with its programs and progress over the years are acutely aware of the need for safety during construction of every one of these aircraft. As a result, the organization has appointed area designees to assist, free of charge, with members' problems arising during fabrication or flight test of any new homebuilt airplane. In addition, design and construction forums stressing safety and quality are conducted during the annual convention held each year at Oshkosh, Wisconsin. Probably the greatest assemblage of lightplane enthusiasts in the world, "Oshkosh," as this midwestern Mecca is called, attracts over 300,000 people during an 8-day period usually beginning the last Saturday in July. With over 45,000 active members of whom some 35 to 50 percent make the trip to Oshkosh in any one year, the how and why of flight safety certainly should be carried back home for the education of all.

Despite all safety messages many rugged individualists still intend to learn to fly in the airplane they are building. Some of these aircraft are quite sensitive in pitch because of their light weight and short coupled design, requiring considerable pilot skill during takeoff and landing (Ref. 1.3). And to further complicate matters, many aircraft are equipped with modified automobile engines not designed for continuous, high-speed operation—providing an open invitation to powerplant failure and off-field landings by unskilled pilots.

An equally serious safety problem lies in some builders' tendencies to substitute readily available materials for those called out on the design drawings, or to radically modify the configuration without being aware of the potential danger involved. These practices are so prevalent that most designers require

FIGURE 1-5
The Pitts aerobatic biplane may be homebuilt from plans and kits or purchased as a finished airplane.
Pitts Aviation Enterprises Photo

every builder to sign a liability release prior to purchasing a set of drawings. Fortunately, through articles in *Sport Aviation,* the excellent EAA journal, and design meetings open to the membership at Oshkosh, the message is driven home: "Do not alter the materials or design configuration in any manner without first receiving written approval from the designer."

While apparently there is no exact record, estimates of homebuilts under construction at any one time seem to vary from 5000 to 10,000 aircraft. Basing my conclusions upon observation of EAA chapter activity around the country, I expect the actual number to be nearer 3000 to 5000 aircraft. (These totals exclude the reported 4000 deposit orders for factory-built Bede BD-5 aircraft, which now appear gone forever.) At any rate, the number of homebuilt aircraft under construction is large and thus the safety exposure is enormous. Of the homebuilts completed, aircraft flown by competent pilots have the most successful records and most frequent usage.

To the extent possible, the EAA maintains files on all homebuilt accidents with particular emphasis placed upon aerobatic mishaps involving experimental aircraft (homebuilt or modified production types) as well as the more numerous certified models. The information in Figure 1-6, kindly supplied by Mr.

ACCIDENTS BY AIRCRAFT TYPE		
Amateur built (homebuilt) aircraft		100
Other civil aircraft		381
	Total	481
Fatal aerobatic accidents		355
Total fatalities		511
Percentage of accidents that were fatal		73.8

ACCIDENTS BY YEAR			
Year	Number	Year	Number
1977	28	1970	34
1976	31	1969	24
1975	21	1968	34
1974	48	1967	34
1973	39	1966	41
1972	43	1965	31
1971	33	1964	40

ACCIDENTS BY CAUSE		
	Number	Percent
Pilot error	371	77
Structural	41	9
Mechanical	38	8
Other	31	6
Total	481	100

FIGURE 1-6
Aerobatic accidents (1964–1977).

Ben Owen, executive director of aviation safety for EAA, covers the 14-year period from 1964 through 1977.

While the total number of aerobatic accidents was relatively small over the years, the resulting fatalities were high. Note also that the pilot error percentage closely equals the Figure 1-1 value for 1948, and that over three-fourths of these accidents happened with other than homebuilt aircraft—that is, with certified types.

Although various conclusions might be drawn from this comparison, the data seem to indicate that many aerobatic aircraft are more nearly designed to 1948 strength and quality standards than to our more rigorous current requirements. And so these aircraft, most of which are recently manufactured copies of pre-World War II designs, are less forgiving of heavy-handed pilots and aerobatic students.

From this we can logically conclude the need for a modern aerobatic airplane having higher load factors and more sophisticated control systems than were used for older designs. The new Sequoia S300 (Figure 1-7) is typical of this next generation of rugged, high-performance aircraft. They will be capable of aerobatic flight at reduced gross weight with accompanying high allowable load factors.

Although the particular EAA data of Figure 1-6 are admittedly limited to aerobatic accidents, they indicate that homebuilt aircraft are involved in a small percentage of all reported accidents—even in the aerobatic category where a greater percentage of aircraft are homebuilt than would be found in the personal aircraft group.

In fact, with approximately 185,000 single-engine piston planes licensed in the United States today, only about 7000, or just under 4 percent, are homebuilts. As a result of the EAA safety effort, homebuilt airplanes are only involved in about 4.5 percent of all small aircraft general aviation accidents, totaling 171 out of 3770 for 1976 (Ref. 1.4). This value is similar to the percentage of registered homebuilt aircraft noted above, and represents an operational record approaching that of certified production aircraft. And a record in which the EAA takes justified pride while steadily working to lower the totals year after year.

*   *   *

The balance of our book will be concerned with many ways to reduce accidents, so you may now rest assured that the dry discussion concerning basic regulations, procedures, and legal terms is out of the way. Let us next see where changes can be made and what they might be.

REFERENCES   1.1   (a)   *A Statistical Analysis of Non-Air Carrier Aircraft Accidents for the year 1948.* Civil Aeronautics Board Report (unnumbered).
     (b)   *Annual Review of Aircraft Accident Data—U.S. General Aviation Calendar Year 1970.* National Transportation Safety Board Report NTSB-ARG-74-1.
     (c)   *Annual Review . . . 1971.* Report NTSB-ARG-74-2.
     (d)   *Annual Review . . . 1975.* Report NTSB-ARG-77-1.
     (e)   *Annual Review . . . 1976.* Report NTSB-ARG-78-1.
     (f)   *Annual Review . . . 1977.* Report NTSB-ARG-78-2.

1.2   Rules Pertaining to the Notification and Reporting of Aircraft Accidents or Incidents . . . , NTSB Regulation Part 830, paragraph 830.2.

1.3   (a)   John W. Olcott, Bede Turns It Loose, *Flying,* October 1974.
     (b)   *Aviation Consumer,* vol. VI, no. 10, May 1976, p. 3.
     (c)   Leroy Cook, Bede BD-5, *Private Pilot,* October 1975.

1.4   EAA accident study data, Nov. 9, 1977.

OPPORTUNITIES FOR IMPROVEMENT

When the types of accidents listed in the tables of Chapter 1 are analyzed in detail, it is possible to determine the primary causes of those accidents over which aircraft design, air traffic control, and flight personnel can exercise some degree of ingenuity to reduce pilot error and improve flight safety.

ACCIDENT CAUSES

Although these possibilities are reviewed and studied at some length in the following chapters, a summary of the findings will provide useful background for the balance of this book. The following list is not intended to be complete; nor are the items necessarily listed in any critical order; however, these causes of accidents may be eliminated through improved design and flight procedures, and therefore each one is of primary importance:

1. Low-speed flight accompanied by stall/spin or hard landing (elevator control mismanagement)

2. Marginal control in crosswind conditions

3. Retractable landing gear

4. Weak landing gear, particularly for nosewheel attachment and side loads on main gear

5. Structures which are not crash resistant

6. Aerobatic load factors which are too low

7. Fire after impact due to fuel location and active electrical system

8. Loading outside approved c.g. limits

9. Pilot fatigue

10. Fuel system mismanagement

11. Powerplant controls mismanagement

12. Powerplant icing

13. Propeller torque accompanying sudden changes in power

14. Engine and propeller failures

15. Poor visibility leading to ground and air collisions

16. Poor aircraft maintenance

17. Airport location resulting in dangerous approaches—valleys, trees, high obstructions, wires, etc.

18. Confusing runway designations

19. Difficulty in determining wind direction

20. Airport surface in poor condition

21. Poor or no communications with airways and airport personnel

22. Confusing and overly complex enroute procedures

23. Improper or no preflight inspection of aircraft

24. Density altitude effects—takeoff, climb, and landing

25. Overloading aircraft for takeoff

26. Improper technique for handling ice, snow, slush, and heavy rain on runway

27. Improper technique on water

28. Improper approach technique—VFR and IFR

29. VFR pilots flying into IFR conditions

30. Wake turbulence, gusts, and wind shear problems

The first 15 items are detail design features which could have received more attention in the engineering office when existing airplane configurations were established. Solutions to these and other design problems will be discussed in Chapters 3 through 7; every improvement considered could have been incorporated in all small aircraft built over the past 30 to 40 years. Their initial application would not require advanced development or press state-of-the-art engineering capability, but such features could stress corporation pocketbooks if added after a design received approval without them.

For personal aircraft the greatest safety contribution would be the incorporation of more positive stall warning or complete stall prevention features in our basic aircraft designs—*because an airplane that will not stall cannot spin*. Various design approaches and configurations that would virtually eliminate stall/spin accidents are presented in Chapters 3 and 4. Once these possibilities have been explored, it seems logical to inquire why our small personal aircraft have not been stall-free for many years. Probably the most realistic answer is that to accomplish this goal manufacturers would be faced with the cost of developing new aircraft, plus the equally burdensome financial risk and time delay required for FAA approval of any unconventional airplane.

DESIGN CRITERIA FAA aircraft design regulations provide minimum standards and requirements, which may not be as stringent or as safe as desired for many types of operation conducted at different levels of flying ability. Once an airplane is approved by the FAA, subsequent design changes are both time-consuming and costly—and so are avoided like the plague unless found absolutely necessary for flight safety as mandated by service problems or Airworthiness Directive (AD) notices. (ADs are mandatory service or repair orders issued by the FAA against aircraft and powerplant items found to have operational or structural defects in service.)

FAA basic design criteria apply to conventional aircraft. This is clearly stated in the old Civil Aeronautics Manual 3 (CAM 3), Appendix A, simplified analysis for single-engine aircraft, and has been similarly carried over into current Federal Air Regulations (FAR) Part 23 requirements—with the volunteered interpretation that "a Piper Cub is a standard conventional airplane." And when the going gets heated during a design disagreement, FAA personnel fall back to that standard to establish a beachhead. Possibly this is desirable from a conservative design point of view, but such practice may actually contribute to flight fatalities by prolonging old design configurations rather than encouraging new, spinproof types.

Fortunately, our major manufacturers are moving ever closer to spin-resistant aircraft through gradual model changes, providing gentle stall characteristics and including stall warning indicators as standard equipment on production units. But the term remains *spin-resistant* when it really should be *spin-proof* for aircraft flown infrequently by private pilots. In that regard, please note that these comments apply primarily to aircraft flown only 50 to 200 hr per year by an individual private pilot; airplanes flown professionally or for aerobatics are exposed to a different brand of flying—supposedly being handled most capably by the professionals and equally well during aerobatic maneuvers requiring stalls.

The powerplant-related items result to a large degree from deficiencies which could be eliminated at the source. With commercial aircraft production reaching 15,000 units annually in the United States alone, volume should permit the use of more sophisticated and automatic powerplant controls at very little increase in basic price. In an industry so heavily labor intensive (that is, requiring so many expensive labor hours to build each airplane), the additional cost of even as much as $1000 worth of purchased basic equipment would add only about 5 percent to the final list price (Ref. 2.1). And greater safety is certainly worth that expense; every year the savings in hull insurance alone would return a large percentage of the increased initial cost, to say nothing of the additional safety provided. With such potential in mind, powerplant improvements are reviewed in Chapter 5.

COLLISIONS   Because collision accidents are frequently the result of other problems, the total number remains high year after year. Solutions to item 15 along with related causes are covered in Chapters 6 and 7. We again find some new design approach essential to collision reduction, although it is doubtful whether the aircraft configurations and equipment required will ever see production unless supported by federal research and development funds. However, as a beginning, much could be done to improve visibility on the ground and in the air.

FLIGHT OPERATIONS   While flight procedures and techniques are not normally considered to be within the aircraft design sphere of influence, the subject has been included in this book because design and flight characteristics are basically interrelated. Furthermore, such a large percentage of accidents stem from improper flight operation that items 23 through 30 represent the largest number of causes of any accident group. In Chapters 8 through 10 we will discuss ways some of these problems can be improved or eliminated through flight technique, new equipment, and aircraft design features. Interest will again be directed toward the serious but occasional IFR and VFR pilot, rather than purely sport flying, professional, or aerobatic operation. Although weather is always lurking about and seems to defy engineering control, some methods of successful combat will also be covered.

Item 16 is another singular cause of accidents we shall review in Chapter 11. The list of cures runs from maintaining proper weight and balance records through adequate airworthiness inspections, rapid incorporation of AD notes, and proper preflight checks. Since many small chores such as preventative maintenance can be performed by the owner under the supervision of a qualified airframe and powerplant (A&P) or an inspection authorized (IA) mechanic, there is little excuse for any well-appreciated airplane to lack adequate maintenance. After all, aircraft are much like boats; half the pleasure of ownership is taking care of them and making constant small improvements. Such effort is also rewarded by higher resale value when the time comes to buy another airplane.

AIRWAYS AND AIRPORTS   Airway and airport problems create a consistent number of accidents as listed in Figure 1-1, with major causes set forth in items 17 through 22. While not as much as might be desirable can be done to correct many of these items at existing airports, the guidelines and suggestions set forth in Chapter 12 should at least improve present conditions while making future airports considerably safer.

Flight along our airways could be simplified, particularly in the more remote parts of the country, while the techniques of controlled and uncontrolled communication are overdue for revision as discussed in Chapter 12. The net results should be a reduced burden upon pilots who find themselves in "tight" situations or severe IFR weather; however, the changes proposed are probably not suitable for major terminal control areas (TCAs), which are really the domain of professional pilots as far as occasional VFR and IFR private flying is concerned.

EFFECT OF IMPROVEMENTS
The possible effect of reduced accident rates upon general aviation operations is examined in Chapter 13, with the conclusion that accident reduction is mandatory *now* if insurance costs are to remain at present levels. A major design improvement such as a fully controllable stall/spinproof airplane should not only reduce insurance premiums but also open flying to many people unable to obtain adequate insurance coverage with present aircraft. Of course, liability premiums to manufacturers would also be decreased, lowering overhead rates and list prices—so the cost of flying should come down along with the accident rate and operating costs.

Where we can go from here and how to get there is the subject of Part Two, Safety by Design, which includes some representative aircraft configurations incorporating recommended design details and handling characteristics. We will begin this safety review by considering possible improvements in aircraft design features.

REFERENCE
2.1 David B. Thurston, *Design for Flying.* McGraw-Hill Book Company, New York, 1978, chap. 12, pp. 174–175.

PART TWO

SAFETY BY DESIGN

STALL PREVENTION

WHAT CAUSES SPINS

If an airplane is flown absolutely perfectly, it will not stall. Even if the difference between maximum speed and stalling speed is as small as 5 mph, an exceptional pilot can keep the airplane from stalling—provided, of course, there is no turbulence or gust action. That's why pioneer aviators usually flew early in the morning and in the evening, or whenever there was virtually no wind or air current activity. The ones who lived were natural pilots, many teaching themselves to fly. Even if they did not know why, the survivors fully realized their aircraft lacked a safe margin between flying and stalling speeds as well as sufficient control response to overcome angle of attack changes in gusty air.

But relatively few pilots in the course of aviation history have been so gifted. In fact, the levels of attitude sensitivity and control possessed by the more noted pioneers remain extremely rare today and probably could not be found in 1 percent of the licensed civilian pilot population. This means that if flying is to be made safer for most licensed pilots, something must be done to reduce or eliminate stall-induced accidents.

Preventing the stall is most important because as previously noted an airplane that does not or cannot stall will not spin; wing lift must be assymetrical for an airplane to begin spinning. When one panel loses lift at a high angle of attack, which means it has stalled, the forward lift component of the opposite active or partially active panel plus the drag of the stalled panel starts the rotation entry into a spin. The basic initial forces leading into the spin condition are shown in Figure 3-1.

By this time the horizontal tail is also at least partially stalled due to the high angle of attack of the airplane, allowing the wing moment M_{ac}, acting about the aerodynamic center (a.c.) of the active wing, to depress the nose. Thus we are now in a nose-down, circular rotating maneuver commonly called a *spin* — or, more properly, a *tailspin,* since the tail tends to rotate on the outside of the rapid descent path.

27

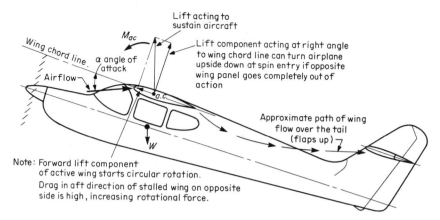

Lift acting to
sustain aircraft

M_{ac}

Lift component acting at right angle
to wing chord line can turn airplane
upside down at spin entry if opposite
wing panel goes completely out of
action

Wing chord line α angle of attack

Airflow

a.c.

Approximate path of wing
flow over the tail
(flaps up)

W

FIGURE 3-1
Conditions acting at spin entry. M_{ac} tends to depress nose.

Note: Forward lift component
of active wing starts circular rotation.
Drag in aft direction of stalled wing on opposite
side is high, increasing rotational force.

Once an airplane is fully into a spin, both wing panels plus the horizontal tail are normally fully stalled. That is why rotation must first be stopped with the rudder to begin recovery. The resulting dive condition restores horizontal tail effectiveness, permitting application of nose-up trim to regain wing lift. However, caution is required to limit the amount of "up" elevator applied; too much nose-up pitch can restall the wing at high speed, causing a secondary spin entry from a lower altitude—which makes matters even worse.

The fact that a lifting wing at high angle of attack has a forward component as shown in Figure 3-1 was not recognized for some time; as a result, many early aircraft failed in flight from wings collapsing forward. Since this action seemed to defy the laws of physics as well as being counter to the airloading felt by pilots sitting out in the breeze, neither these structural failures nor the frequently fatal spins were fully understood until airfoil and flight theory became a bit more advanced around 1915.

Although this has been a rather basic and simplified discussion of spin conditions, it will suffice for our purposes since it identifies the elements of a spin: one wing stalled, one wing at least partially lifting, and a partially or fully stalled horizontal tail. Note from Figure 3-1 that if one wing panel is completely out of action while the opposite side retains a large percentage of lift, which can happen in a tight turn at low speed, it is possible for the airplane to roll over and enter a spin upside down. This is an inverted spin entry, which can be a dangerously uncomfortable attitude if the tail remains stalled, particularly at low altitude such as a crosswind correction on final to recenter the runway. Many approach accidents occur in exactly this manner, frequently amplified by crossed controls at a speed too low to sustain yawed flight.

ACCIDENT CONDITIONS

As evidence of improved stability resulting from more stringent FAA design requirements, the total number of stall, spin, spiral, and mush accidents has been slowly decreasing over the years even though more aircraft are flying. Unfortunately, this classification still accounts for over 10 percent of the first type of accident reported—exceeding 400 wrecks per year accompanied by a high fatality rate, a total much too high to be tolerated.

The stall/spin accident is generally listed as a pilot error situation and probably is in most instances. But in many cases the pilot, absorbed in other demands, does not realize the airplane is slowly going out of control. Basic inattention or being lost in the overcast with only a VFR rating can be so distracting that even a 15 to 20° change in pitch attitude goes unnoticed until the spin windup or steep spiral begins. And concentrating on the elusive approach end of a runway can frequently result in slowly but steadily bringing the nose up to stretch a glide until the flight ends in a stall/spin or by mushing into the ground.

It seems quite obvious that something should be done to make the pending stall so apparent that no amount of distraction could prevent recognition of the number one flight problem, namely: When speed is too slow, down we go.

NTSB investigation has shown that some aircraft designs are much more likely to spin than others and thus have a much higher accident rate. Their most comprehensive study (Ref. 3.1) weighs spin accidents by design type in relation to aircraft flight hours, with the comforting finding that our newer designs are considerably safer than the old standbys so frequently reviewed in glowing nostalgia.

For example, for noncommercial flying the Aeronca 11 series, Piper Cubs and Super Cubs, Cessna 150s, and Taylorcraft B's are rated very prone to stall/spin accidents, while the Cessna 172, 180, and 206 are listed as having a low frequency of stall/spin accidents along with the Piper PA-24 series. The safest of all were found to be the Beechcraft 35/33 models, Cessna 182 and 210, and the Piper PA-28 Cherokee line, giving further proof that the more recent FAA stability requirements are mandating the development of safer aircraft by design. However, it seems possible that the effect of student training in the Piper Cubs and Cessna 150s may not have been thoroughly considered in this study, although the Cherokee is used for training and rated very well.

With regard to training, restoration of spin recovery practice would help to reduce spin accidents. This used to be part of post-solo flight instruction for the private pilot license and resulted in every pilot knowing what a spin was and felt like, and, more important, how to recover from the spin attitude with minimum loss of altitude. Spin training at least would tend to eliminate pilot fear of the spin condition while also teaching early recognition of impending stalls.

Not all stall/spin accidents originate during the cruise or approach phases of flight. Takeoff seems to claim a steady stream of victims during initial climbout. The basic reasons are classified as the failure to obtain/maintain flying speed, improper weight distribution, and overloading the airplane.

As indicated in Figure 1-2, a fairly constant but slightly decreasing percentage of pilot error accidents are attributed to failure to obtain flying speed. Such accidents during takeoff usually occur as a spiral or mush into the ground, or as the beginning of a spin if sufficient altitude has been attained. Behind this type of accident lie density altitude problems, powerplant failures, and poor flying technique.

Stall/spin/mush accidents of this type have been rather consistently distributed over the years with about 25 percent occurring during takeoff, 35 percent during landing (in pattern and final approach), and 40 percent from in-flight operation (aerobatics, buzzing, low passes, etc.). Since a spinning airplane is fully stalled, it descends as a free-falling body. Unless there is considerable altitude to permit recovery, usually in excess of 1000 ft above ground level, the end is quite final; in fact, spin accidents have the highest fatality rate of any type.

POSSIBLE CURES
In view of the above conditions, more should be done to eliminate stall/spin accidents, particularly since we have prevention measures available for the asking.

DESIGN FEATURES
There is a pre-stall warning device on most production aircraft, but it is frequently set at too low an airspeed for gust conditions, has been disconnected, or is just not heard because of concentration on other flight problems. Some commercial transport and business jets have stick shakers working in parallel with aural and visual warnings to alert the most tardy pilot. Other systems go a step further by including a device to push the stick (or wheel) forward when speeds become too low. While it may be properly argued that any mechanical warning system is prone to failure just when needed, this should not be so with dual wiring and modern solid-state electronics.

In addition to lights, buzzers, and stick shakers, there is one seriously undervalued instrument that could solve most of the in-flight speed/stall/spin problems as well as increase flight efficiency if pilots would only use it properly— the angle of attack (AOA) indicator. If this instrument does not receive popular acceptance more rapidly, it may be only a matter of time before it is legislated into the cockpit.

There are other solutions to the stall/spin problem which are offered to designers through use of basic aircraft design features that will (1) make it difficult for a pilot to obtain a high pitch attitude from a trimmed condition, (2) improve spin recovery characteristics, (3) provide stall warning from buffeting action of air passing over the tail, and (4) prevent spinning by making the airplane incapable of stalling. The last approach is particularly desirable because the stall/spin problem is eliminated before it gets to the pilot. As reviewed in Ref. 3.2, proper tail design can provide almost immediate recovery when a spin does occur, while restricting elevator-up travel can prevent wing stall and so preclude spinning since an airplane must stall before it can spin.

FLIGHT PROCEDURES Following the philosophy that a safe design should make any pilot work hard to get into trouble, many aircraft require considerable pilot effort to pitch the nose up or down through a large angle once the airplane has been trimmed for a specific flight condition. In addition to making it difficult for any pilot to get near a stalled attitude without a pair of bulging biceps, comparatively high elevator control forces also prevent changes in pitch attitude with minor movement around inside the cabin in search of charts, clothing, etc.

Aircraft with elevator control forces that steadily increase with g *loading* (total weight on wings divided by airplane weight) are considered to have *positive static stability*. This condition is represented by the typical elevator stick force vs. flight-speed signature of Figure 3-2. Thus, designing for high stick forces during maneuvering flight, which includes turns and landing flare, increases flight safety by making an airplane difficult to stall.

Late models of the Cessna 177 Cardinal have an excellent buildup in stick force during flare (leading into a stall) when trimmed for the approach, although things get a bit busy if a go-around is necessary after the flaps have been dumped while the airplane is still trimmed nose-up for the approach; if this happens, forward elevator pressure may be required to prevent climbing into a stalled condition. The alternative requires carrying a heavy untrimmed elevator load during the approach, but this can cause an airplane to dive if the control is accidentally released during final. Thus, too much static stability also has its problems.

The use of an airplane's natural aerodynamic characteristics can provide stall warning by making the horizontal tail shake from the buffeting action of turbulent air flowing off the wing root area. A desirable wing stall will start at the root trailing edge alongside the fuselage and gradually spread forward and

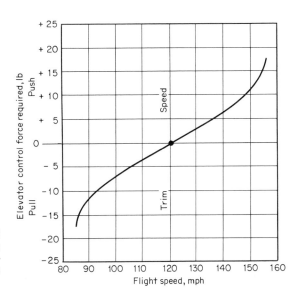

FIGURE 3-2
Elevator control force variation with flight speed. Stable airplane trimmed for 120-mph cruising flight.

outboard as the pitch attitude (wing angle of attack) is increased. This action is diagramed in Figure 3-3.

When turbulent flow impinges upon the horizontal tail, the structure is disturbed sufficiently to set up a pre-stall warning vibration that can be felt in the control system and aft fuselage. Some aircraft are fitted with leading edge airflow spoilers to augment flow breakaway from the wing root, but this practice tends to reduce wing lift all the way forward to the leading edge just when maximum help is needed at landing. Leaving the wing root trailing edge partially unfilleted is a more effective and positive means of obtaining root flow breakdown at higher angles of attack, and need not increase cruise drag.

The no-stall solution to spinning was developed to a high degree by Fred Weick and his Ercoupe design of the late 1930s and postwar period. Many of these two-place aircraft are flying today, with quite a few retaining the original two-control feature of separate elevator plus interconnected rudder and ailerons. With restricted elevator-up travel and no rudder pedals to cross control, this airplane was spinproof—but it could be pulled up into an abrupt pitch attitude requiring some 200 to 300 ft of ground clearance for recovery to normal flight.

I recall an Ercoupe incident during 1946 at the Hicksville Aviation Country Club on Long Island, New York (now part of Levittown), when such a demonstration maneuver nearly cost the life of one of our Grumman test pilots; his recovery was a scant 50 ft above the surface. Thus, even with restricted elevator, inertia forces can produce an accelerated and steep pitch-up requiring considerable airspace for recovery if the stall condition can be approached too closely.

SUMMARY

In summary, any one of the following design and flight procedures would be of assistance in reducing the frequency of stall/spin accidents:

FIGURE 3-3
The use of controlled, progressive wing stall to provide stall warning through tail buffeting. Flaps-up condition. (a) Root stall begins and weak turbulence flows over the horizontal tail (about 10 to 12° AOA). (b) Wing stall is well advanced with strong turbulent flow producing tail buffeting. Note ailerons still effective (about 14 to 15° AOA).

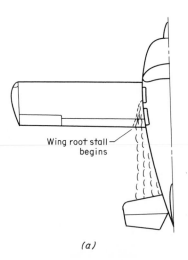

Wing root stall
begins

(a)

Wing stall moving
forward and
outboard

(b)

1. Thorough pilot training plus semiannual or annual proficiency checks for the type of airplane flown

2. Stall warning indicator (aural and visual)

3. AOA indicator

4. Stick shaker (warns of impending stall)

5. Stick pusher (moves stick forward to prevent stall)

6. Buffeting airflow over horizontal tail caused by wing root stall

7. Increasing elevator control forces with increasing angle of attack (stable airplane with stick forces increasing with g loading)

8. Improved spin recovery characteristics through proper tail surface design

9. Airplane designed to be incapable of spinning—two-control type which eliminates rudder pedals (and possibility of crossed controls)

10. Development of a density altitude–sensing takeoff speed indicator

11. Development of a simplified windshield-type or head-up display (HUD) for airspeed and AOA readout similar to military equipment. (A *head-up display* presents flight data on an inclined, clear glass screen located in the pilot's line of vision—hence, information can be noted with the pilot's head up in normal flight position.)

With all these options possible, it seems strange they have not been combined to eliminate the stall/spin problem. However, the important point is that although stall/spin accidents are primarily caused by pilot error, we presently have many ways to overcome this failing through proper design.

Considering the average levels of flight technique and pilot proficiency, let us now discuss aircraft that may be stalled but could be prevented from doing so through the use of existing or special flight equipment within current state-of-the-art technology, and then go on to study basic aircraft design features capable of developing a fully controllable, stallproof airplane.

EXISTING EQUIPMENT

First, we will consider aircraft that stall and thus can spin.

STALL WARNING INDICATOR

Everyone who flies a fairly modern airplane is familiar with the stall warning indicator manufactured by Safe Flight Instruments of White Plains, New York. A small vane usually mounted on the left wing leading edge signals the presence of a stall warning system that senses airflow (local angle of attack) changes over the wing. This vane is positioned by flight test to move upward at a critical airflow shift along the wing leading edge that will provide a warning some 5 to 10 mi above stalling speed. When this flow change occurs, upward movement of the vane actuates a microswitch, which in turn energizes a red warning light and buzzer mounted on the instrument panel. Almost standard equipment in today's aircraft, this pre-stall warning system and circuit are shown in Figure 3-4.

Claiming that the buzzer can be annoying during slow-speed flight in turbulence, many owners disconnect this part of the system. Unfortunately, anyone noting a stall warning vane installation during preflight inspection will subsequently expect to hear a buzzer as stall speed approaches; if no buzzer sounds and the red warning light is not noted, an airplane can be slowed into a full stall with an operative stall warning system aboard. Quite obviously, it is poor practice to disconnect the stall warning buzzer, except during experimental slow-flight tests performed by a professional test pilot. And those occurrences are rare.

Not so obvious and generally ignored is the precaution of checking the stall warning buzzer and light during preflight inspection by turning on the master switch and moving the stall warning vane up and down a few times to find whether the buzzer goes on and off in the cabin. Assumedly, the light also comes on—but the buzzer is more important because climb-out or approach conditions near a developing stall may preclude glancing at the instrument panel during flight. In view of this, possibly the best place for the stall warning light is right on top of the instrument cowl near the pilot's line of sight. In this position the red light should be noted if the buzzer or its part of the warning circuit fail for any reason.

AOA INDICATOR   The AOA indicator should be mentioned here in regard to its stall prevention features. By indicating variations from the best climb or approach trim angle, as determined by flight test and set for each different aircraft model, an AOA instrument provides information to correct speed changes due to pilot technique,

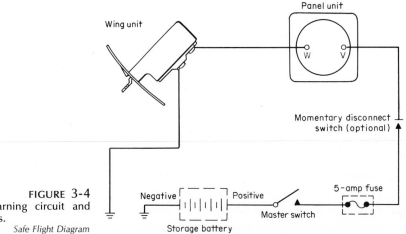

FIGURE 3-4
Pre-stall warning circuit and components.
Safe Flight Diagram

local turbulence, or wind shear—and in time to prevent stalling. Mounted within the pilot's line of sight to permit viewing while the pilot is looking outside the airplane, the AOA indicator may be combined with a stall warning system buzzer and light for maximum possible safety—a combination that would surely prevent many pilot error stall/spin accidents resulting from "failure to obtain/maintain flying speed" as listed in Figure 1-2.

CONTROL STICK SHAKER   For those hard-to-convince cases there are control stick (wheel) shakers available that act a few miles above stall speed to warn otherwise occupied pilots that trouble is fast approaching. Activated by a sensing vane similar to that used for AOA indication, the stick shaker consists of an eccentric weight attached to the control stick or wheel system to provide a fore and aft vibration that cannot be overlooked; this warning is given in time for the pilot to safely reduce pitch attitude, thereby increasing the speed margin above stall for a given power setting.

STALL WARNING SYSTEM   Aircraft with a very sharp or critical stall break may require the basic stall warning buzzer/light system, a stick shaker, and a stick pusher. The stick pusher does just that; it rapidly pushes the stick or wheel forward to relieve high angles of attack before a deep stall or unrecoverable attitude can develop. The corrective signal is again developed by an external vane. Since this device is fairly heavy, expensive, and rather difficult to install, it is not really intended for small general aviation aircraft. However, this equipment along with a stick shaker is used in some business jets and on many different models of commercial transport aircraft.

Even though our newly designed "safe airplane" will be stallproof and spinproof, the simple buzzer/light stall warning indicator will be used to indicate undesirably low operating speeds, possibly bordering upon marginal crosswind control capability or excessive sink rate. In addition, the AOA indicator will be included for proper climb-out attitude and speed range, to simplify approach setup, and to aid possible balked landing or missed approach procedures. The information and assistance offered by these two instruments will prove helpful while also increasing safety throughout the entire flight profile.

NEW EQUIPMENT   The accident category listed as "failed to obtain/maintain flying speed" is filled with pilots who attempted a takeoff but never succeeded. Failure may have been due to density altitude, attempting to break ground too soon, overloading, improper loading (outside the approved c.g. envelope), or a combination of these causes. But the end result is always the same; attempting flight below minimum flying speed produces an accident. Once forward momentum has been dissipated in a short climb, the forces of gravity take over in an uncontrolled crash—with the ground, trees, buildings, or any other obstacles along the intended flight path.

TAKEOFF SPEED INDICATOR

Although climb-out stalls may be prevented by AOA and stall warning indicators, takeoff accidents occur while accelerating prior to attaining flight speed and so cannot be indicated by either existing instrument. The takeoff speed, or flyaway, indicator proposed in Chapter 4 will provide a "go" signal when the airplane reaches safe takeoff speed, regardless of density altitude conditions. Since it will be developed to sense variations in pressure and temperature by operating as a dynamic pressure indicator (see page 50), the safe flyaway signal can also be made controllable for variations in airplane takeoff weight, permitting optimum performance based upon actual operating weight. Again, as a reminder, neither overloading nor operating outside approved c.g. limits are curable by instruments, but either will soon "cure" the pilot who follows such practices.

HEAD-UP DISPLAY (HUD)

Use of the HUD system will further improve operational safety for all flight conditions, including VFR and IFR weather. By presenting airspeed, altitude, AOA, rate of climb (ROC), and omni tracking on one "see-through" panel, it is possible to have this essential information constantly available during critical takeoffs and approaches. The data will also be there during noncritical moments, adding a sense of stall-free security to every takeoff, landing pattern, and approach operation. With a clear display of speed margins above stall presented within the normal field of vision, use of the HUD indicator should materially reduce the number of stall/spin accidents resulting from failure to maintain flying speed.

Because even a stallproof airplane can be "hauled off" before it is ready to fly, and considering that the HUD system as outlined provides a coordinated data source for optimum performance regardless of aircraft stall characteristics, the flyaway speed and HUD indicators will be included as part of our "safe airplane" basic equipment list—as soon as they become available for small general aviation aircraft.

But believing the best engineering approach is that of eliminating trouble at the source, let us now consider basic design features intended to warn of or prevent stalls—and so eliminate spin accidents. After these fundamentals have been discussed, we shall look at aircraft configurations providing the desired level of safety.

STALL PREVENTION FEATURES

Beginning with basic design qualities intended to warn of an impending stall, possibly the best safety protection is to have high stick forces accompany speed changes from trimmed flight.

Note that even with adequate longitudinal stability as indicated by rapid

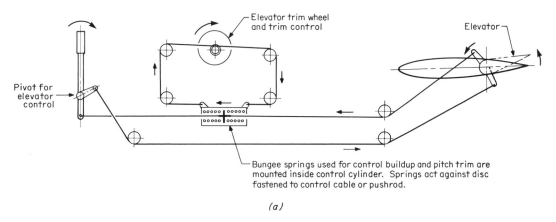

(a)

FIGURE 3-5
Bungee spring system providing pitch trim control and increased elevator load with
stick displacement. (a) Bungee
used for elevator trim and increased control force with
stick displacement. Arrows
show nose-up trim movement.
(b) Bungee for control force
with separate cables for elevator trim tab control system.

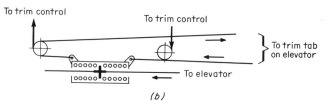

(b)

pitch damping and increasing stick forces with changes from trimmed *cruising*
speed and attitude (Figure 3-2), it is still possible to have very small increases in
stick force when varying from trimmed attitude at *low speeds*. So when pilot
attention is called to other matters, there may be slight or no warning given by
elevator control force buildup at the wheel or stick.

BUNGEE AND
BOBWEIGHT SYSTEMS

The above condition can be overcome by designing load producing springs
into the elevator control system, a form of poor man's fly-by-wire control feel,
also known as a *bungee system*. As shown by Figure 3-5, load can best be applied by movable springs capable of displacing the elevator for pitch trim
throughout the required flight speed range from 1.3 stall speed to red-line never
exceed speed V_{NE}. By having this spring load system move the elevator when
pitch trim changes are required, the springs are constantly in action. For example, the forward spring is ready to increase stick load when up elevator is applied to increase nose-up trim (approaching the stall attitude), while the aft
spring is also right there to increase nose-down control forces whenever the
stick is moved forward.

But there are a few disadvantages with this arrangement. For starters, the
FAA now requires that all certified aircraft have a pitch trim system capable of
controlling the airplane if the primary elevator control system fails. This means
a completely separate pitch trim control system. While I have never heard of

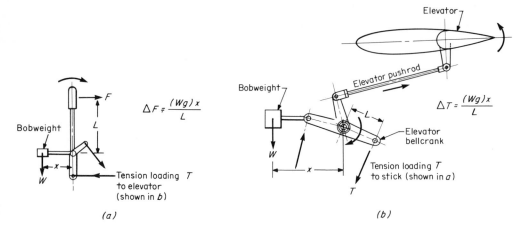

$$\Delta F = \frac{(Wg)x}{L}$$

$$\Delta T = \frac{(Wg)x}{L}$$

(a)   *(b)*

FIGURE 3-6
Bobweight locations used to increase control forces with *g* loading. Either position may be used depending on aircraft weight and balance limitations. (*a*) Bobweight at stick. Acceleration *g* loading acting on bobweight *W* increases control force *F* as *g* loading increases. (*b*) Bobweight in control system. Acceleration *g* loading acting on bobweight *W* increases load *T* in control system, which in turn increases control force *F* at the stick.

any elevator control failure, I presume this could happen through poor maintenance or assembly. In order to satisfy this requirement, it is necessary to have the elevator load bungee moved the same distance by the separate elevator trim tab control that the basic elevator control cable or pushrod is displaced by the trimmed elevator. If this is not done, the load in the bungee spring system could oppose and overcome the trim tab loading and so prevent desired trimming displacement of the elevator surface. However, it is not difficult to move the bungee spring unit with the elevator primary control cables or pushrod if it is connected into the elevator control and trim tab systems as shown by Figure 3-5(*b*). Since they are not restricted by FAA trim requirements, homebuilt or experimental aircraft can obtain desired control force buildup and elevator trim action simply by displacing the bungee to move the elevator to the desired trim position as shown by Figure 3-5(*a*).

As a second item, FAA regulations require that any airplane having springs in the control systems must be dived to design dive speed V_D with the springs connected and also disconnected. This is done while the control surface is rapidly displaced a few degrees plus or minus to demonstrate freedom from flutter. These flight tests must be approached cautiously with the test speed gradually increased in increments of about 5 mph to V_D, hopefully without incident or difficulty of any kind.

A third problem arises with bungee trimming devices used on higher-performance aircraft in that considerable spring force may be required to provide the elevator displacement necessary for pitch trim limits—particularly at gross weight forward c.g. loading, flaps fully down, through the range of minimum to maximum flap operating speed. The considerable nose-down movement produced may result in large, heavy, and powerful trimming springs that will simply overpower the elevator control system. If this design problem appears certain, use elevator trim tabs to overcome pitching moments [with a suitable bungee as shown in Figure 3-5(*b*) if necessary to increase elevator control forces].

Bobweight systems have also been used effectively to increase control forces. This approach essentially consists of a weight cantilevered from the control system in such a manner that under g loading pilot control force must be increased to maintain control surface displacement. Typical types of bobweight attachment are shown in Figure 3-6 for use in an elevator system.

This method of control force buildup normally applies to higher-speed aircraft and is intended to prevent the occurrence of high-g stalls during pull-ups at high speed. By their design, bobweights would have little effect in increasing control forces for low-speed flight near the stall where the loading is only 1 g or slightly higher; thus during low-speed flight they are not a viable means for signaling an approaching stall or preventing stalls. But because they are effective in preventing high-speed, high-g stalls and subsequent violent spins, bobweights are frequently used on aerobatic and other highly maneuverable aircraft.

TAIL BUFFETING   A constant and positive form of stall warning is realized when turbulent air from the wing root area flows back over the horizontal tail as shown in Figure 3-2. To be most effective, flow should remain smooth in all normal flight conditions, with some minor turbulence beginning to gently shake the tail around 10 to 12 mph above stalling speed. As the airplane angle of attack increases, so should the turbulent flow and tail buffeting until a full stall finally develops.

With this feature the beginning of tail shake signals the approaching stall, giving ample warning to lower the nose with about a 15 percent speed margin above stall. For an airplane normally stalling at 60 mph, the onset of tail buffeting at 70 mph provides a warning margin of $10/60 = 0.167$, or 16.7 percent above stalling speed. While this may not be a sufficiently high approach speed or stall margin for gusty conditions, when 30 percent is usually desirable, it certainly is acceptable as a lower limit and better than having no stall warning at all.

Initiating tail buffet at a greater percentage above stall could result in undesirable tail vibration during sustained low-speed climbs or STOL (short takeoff and landing) approaches, so keeping an acceptably low margin between initial buffeting and a fully developed stall condition is preferred design practice.

Now, the question is how to accomplish this? At the higher angles of attack leading into the stall, the horizontal tail must be located within the wing wake as shown in Figure 3-1. Note also from this diagram that a T tail may not be in this high angle of attack wake if used on a low-wing airplane. This would preclude tail buffeting as a positive stall warning indicator, as demonstrated by some of the new generation of T-tail aircraft. However, a low- or shoulder-wing configuration with a low- or mid-fin-located horizontal tail should provide good buffet pre-stall warning as shown by the Sportavia (of Germany) model RS180, the Rallye Series (see Figure 7-4), or the Rockwell 112B of Figure 3-7.

FIGURE 3-7
Rockwell Commander 112B showing raised horizontal tail position.
Rockwell International—General Aviation Division Photo

T-TAIL DESIGN

Since we are discussing horizontal tail surface effect upon pre-stall warning, this is also a good time to discuss the effect of vertical tail design upon stall and spin recovery. As shown by Figure 3-8, the more exposed the rudder surface, the more control response available to correct wing drop and prevent a stall becoming a spin, and the more rapid the spin recovery when required. If the rudder is not sufficiently effective at high angles of attack it may be impossible to prevent a stall from becoming an immediate spin. And when the spin develops it will be impossible to stop rotation and enter the dive attitude necessary for full recovery.

As discussed at greater length in Ref. 3.2, the T tail is working during all phases of flight including control into the stall, spin recovery, and crosswind

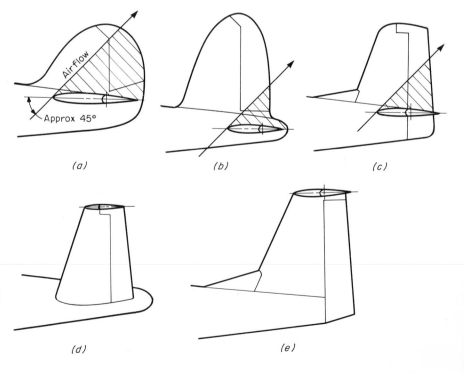

FIGURE 3-8
The effect of horizontal location upon rudder response during spin recovery. (*a*) An early type of rudder design subject to blanketing by the horizontal tail. (*b*) Horizontal tail location for improved rudder effectiveness during spin recovery. (*c*) Rudder extended below the horizontal tail for increased effectiveness during spin recovery. (*d*) Bonanza-type V-tail design provides fully effective tail surfaces during spin recovery (inverted V tail is similar). (*e*) The T-tail configuration has the fin and rudder completely effective with no blanketing.

operation. So if a spinnable airplane is desired, this tail configuration is the safest design possible. But it does present some problems in the form of increased weight, control system complexity, and fuselage loading—plus possible loss of pre-stall buffeting as a warning feature. In view of this, consideration should be given to the use of a full rudder with mid-fin-located horizontal tail for low-wing aircraft. Shoulder-wing and high-wing designs can probably retain pre-stall buffeting with a T tail, while also realizing the excellent rudder control safety advantages offered by this constantly effective vertical surface.

LOCATION OF C.g.   Accidents caused by overloading are considerably more common than those caused by loading behind the aft c.g. limit at allowable flight weight. However, the two conditions frequently combine when overloading occurs. Although by no means so numerous as powerplant-related accidents, overload and aft c.g. causes are frequently buried in statistics listed as "misjudged distance, speed, and altitude" or "inadequate preflight preparation and/or planning," etc. As such, this type of accident is fully attributable to basic pilot error, frequently terminating in a stall/spin collision with the earth or other solid object.

But need this be so? By designing aircraft that could not be loaded outside demonstrated c.g. envelope limits at any weight up to certified gross, we eliminate the opportunity or potential for accidents due to improper loading. This possibility will be thoroughly reviewed in Chapter 6 and so will not be presented here. However, a typical configuration is shown in Figure 6-18, with similar safety offered by other designs presented in Chapter 7.

A new airplane configuration or family of aircraft is really required to take advantage of this basic safety feature. Quite likely the elimination of beyond-limits aft c.g. loading would materially reduce the number of accidents attributed to pilot error resulting from: improper preflight preparation, failure to attain flying speed, collisions with just about anything, stall/mush impacts, nose-overs, and hard landings. This is a sizable list for potential safety improvement, warranting the attention of aircraft designers and manufacturers. A whole new field of users awaits the introduction of basically safe aircraft; and eliminating the possibility of excessive aft c.g. loading would be a step toward realizing that market.

REDUCING TRIM   Variations in pitch trim with power and flap changes produce another source of
CHANGES   stall/spin accidents. This problem is particularly critical when an airplane is trimmed for a "power off," flaps-down approach. If a balked landing or missed approach becomes necessary, it is possible to experience a rapid nose-up pitch change with power application, leading into a power stall and spin from low altitude. For some older aircraft, use of full power may also induce sudden and severe nose-up pitch when trimmed for "power off" final approach. When a

nose-up attitude is also combined with a slip or extreme drift correction at low speed, it is even easier to terminate the landing via a stall/spin maneuver entered without warning. This situation unfortunately occurs quite a few times every year, and usually with fatal results.

The effect of propeller torque produced by sudden corrective application of power at low speed can also be the final touch necessary to start a spin to the left. This would not occur at correct approach speed nor with proper application of right rudder; but rather than provide such pilot error accident potential, why not offset the thrust line to the right to partially or completely eliminate the left turn effect accompanying high thrust values?

While it may not be possible to completely design out trim changes with power and flap settings, there are a few fundamentals that will assist in attaining this desirable handling characteristic. For conventional low-, mid-, or high-wing aircraft it is helpful to have the longitudinal thrust line pass as near the normal gross weight center of gravity as possible. That is, in side view the thrust line should desirably pass through the c.g. In addition, some adjustment of thrust line relative to the wing angle of incidence may be desirable to compensate for trim changes with power or wing drag increases with flap extension.

For example, lowering flaps full down on a high-wing airplane, "power low" or "power off," may result in a nose-up pitch change because wing drag with flaps down is greater than with flaps retracted. And since the flap drag of a high-wing airplane is acting aft and above the c.g., the nose will pitch up. To correct this, use 1° or 2° of nose-down thrust to place the thrust line slightly above the c.g. as shown on Figure 3-9. Because this offset will only be correct for one flap position, it should be calculated for the most critical, which is usually flaps full down during transition from "power off" to full power.

Typically, things are not this simple. When the flaps are lowered, the wing section moment M_{ac} will increase over the flapped area due to increased airfoil camber. This nose-down moment may exceed the added flap drag and so produce a net nose-down pitch change, rather than a nose-up attitude. If calculations indicate this may occur, the thrust line should pass through or below the c.g. In this position the thrust added by increased power will tend to balance out the added nose-down wing moment produced as speed increases with power.

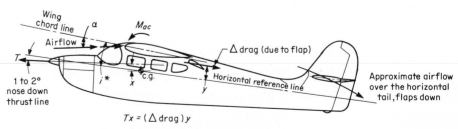

FIGURE 3-9
The effect of offset thrust in correcting trim changes due to flap deflection on a high-wing airplane.

$$Tx = (\Delta drag)y$$

*Note: Angle of incidence i is the angle between the horizontal reference line and the wing chord line.

Low-wing aircraft would be expected to assume a nose-down attitude when lowering flaps because wing and flap drag are both below the c.g., while the wing secion moment M_{ac} also acts to depress the nose. Adding power may further increase nose-down trim, even with the thrust line passing through the c.g., because wing section moment and flap will both increase at the higher speeds accompanying increased power. To correct this undesirable trim change, some aircraft interconnect an elevator or stabilizer adjusting mechanism with flap operation to provide nose-up trim when the flaps are lowered.

This complication may not be necessary and should not be considered unless found mandatory through evaluation of handling characteristics in the approach configuration. Many low-wing aircraft, including the Piper Cherokee series, actually experience a slight "ballooning" or nose-up trim change when flaps are applied. This effect is presumably due to a decreased or more negative angle of wing downwash flow over the horizontal tail caused by flap-deflected airflow from the wings. This flow change in turn produces less up load or more down load on the horizontal tail, depending upon the trim conditions prior to flap operation (see Figure 3-9).

Either way, such a change in the balance of forces tends to produce a slight nose-up trim attitude with flap deflection. It is important that this condition is not amplified by application of power, which could result in a "power on" stall and subsequent spin entry.

The important design safety feature we are seeking is minimum trim change with power regardless of the airplane configuration; this means for any trim condition, including gear and flaps down or up, and at any point within the design c.g. envelope. Demonstration of this capability is required for FAA certification, and in final analysis can only be refined to an acceptable level through flight test evaluation and modification as necessary.

As an equally important safety feature, and rather as a corollary, it is desirable to have minimum trim change with flap deflection, regardless of power setting. If necessary, this can be realized by interconnecting horizontal tail trim with flap deflection as previously noted for flap plus power changes.

If a design choice becomes necessary between slight nose-down or slight nose-up trim change with power or flap application, the nose-down condition should be favored. This pitch attitude leads to increased speed, moving away from the stall and providing time to retrim the airplane. But to provide inherent handling safety in any airplane, large trim movements in either direction must be avoided by design.

Regardless of the final solution, we want an airplane that experiences little or no trim change with flap deflection or power changes throughout the entire flight profile and, more particularly, a design that does not pitch up into a full stall or spin finale. Current FAA handling criteria are providing aircraft with desirably docile characteristics; but, unfortunately, older production aircraft may require some fast handwork when aborting a trimmed, "power off," full-flaps approach. As a result, application of power during final contributes to a steady series of stall/spin accidents, all of which could be prevented by proper aircraft design.

While this trim discussion has been primarily concerned with conventional aircraft fitted with the customary nose-mounted engine, we should also understand that similar safe longitudinal trim characteristics can be realized with a raised engine mounted on T-tail aircraft. The reasons and requirements for this design capability are reviewed in Chapter 6. The point is that acceptable stall-free trim characteristics can be designed into aircraft regardless of engine location, removing another set of opportunities for pilot error through "improper operation of flight controls" and thereby reducing the number of directly related accidents.

NASA MODIFICATIONS   An intermediate design approach to stall prevention has recently been investigated at National Aeronautics and Space Administration (NASA) Langley Research Center. After locally increasing wing leading edge camber by drooping the outer 40 to 50 percent of each panel, flight tests have shown the modified airplane will not enter a spin. Instead, a steep spiral develops following stall, with immediate recovery possible upon release of elevator control back pressure.

Although this appears to be a type of built-in slat design, similar in effect to the movable leading edge slats used successfully by de Havilland, Helio, and others to improve low-speed handling characteristics, rapid spiral descent is still possible if an unexpected stall occurs. So the door remains open to sudden contact with the earth.

In fact, all the stall warning or delaying features discussed so far, with the possible exception of minimizing trim changes with power and flap application, are intended to correct handling effects of basically undesirable aircraft design characteristics. A more sound engineering approach would be that of eliminating any tendency or opportunity for an airplane to stall in the first place. The current level of small aircraft technology will permit this quantum jump in flight safety, and do so while retaining handling characteristics required for crosswind, approach, and STOL performance.

As a matter of interest, the successful Wright glider of 1902, used to develop the configuration for the first powered airplane of 1903, had interconnected wing warping and rudder controls plus a separate elevator control. And so the first powered flight was made with a two-control airplane, although elevator deflection was not restricted to prevent stalling. This combined lateral and directional control system is clearly shown on the excellent replicas at the Wright Brothers National Memorial in Kill Devil Hills, North Carolina. With light wing loading and high drag, these first aircraft had no problem with making steep descents at low speed—so lack of spoiler control was no handicap. At that time staying airborne was the challenge, depending upon ridge wind over the dunes or full-power performance—when either dropped, so did the aircraft.

We should now come full circle back to the original two-control arrangement successfully used on the Wright aircraft. The control system found most responsive to their limited flight experience would also be most easily handled by students and low-time pilots or those suddenly faced with extreme workloads under trying flight conditions.

Let us see how such an airplane can be realized through a proper combination of available design procedures.

THE STALLPROOF AIRPLANE

A two-control method of stall/spin prevention was briefly noted earlier in this chapter, and one possible airplane configuration is shown later in Figure 6-18. The important features of this design approach are: (1) interconnection of aileron and rudder controls to provide coordinated turns while preventing crossed controls and (2) separate elevator travel restricted to prevent wing stall although permitting adequate landing flare throughout the range of c.g. travel. This second item represents a king-size problem that can only be solved by combined design and flight test evaluation; however, the worthwhile end result is an airplane characteristically incapable of spinning.

When elevator-up travel is limited to prevent both slow entry and accelerated stalls, it is essential that trim changes with power or flaps be as minor as possible; in fact, zero pitch change is the ideal goal. So our stallproof airplane requires design coordination between thrust line position, flap deflection, landing gear position (if retractable), and any other flight configuration changes producing moments that will upset or change the trimmed attitude. Although practically any airplane design can be made to fly, only a well-developed one will exhibit the desired minimum level of trim change. And a series of step-by-step changes accompanied by many test flights may be necessary to achieve this degree of refinement.

At this point in our discussion I anticipate some concern about crosswind operation without independent rudder control. When taking off or landing in a crosswind with any conventional three-control airplane, current practice requires a slight bank into the wind to check drift plus top rudder to keep the runway boresighted. And so we have a slight to medium crossed-controls condition during a crosswind takeoff or landing. Too much in the way of crossed controls can, of course, result in a stall/spin from low altitude. Therefore, we must interconnect the aileron and rudder systems to prevent kicking an unstalled wing into a tight turn, since this maneuver could stall the inside panel and start a spin. In fact, a two-control airplane has no separate rudder pedals, only a stick or wheel for elevator and aileron (with interconnected rudder) control—hence the origin of the term *two-control*.

To handle crosswind takeoffs the two-control wheel or stick is displaced toward the wind source—that is, to the left for a crosswind from the left—and takeoff made in a shallow banked climb. If important, once airborne, the bank can be adjusted or converted to crabbed flight to keep the runway centered.

The recommended landing procedure is equally simple. During the final leg of the approach, bank the airplane sufficiently to check drift, level the wings to follow a crabbed track over the ground in line with the runway, and land on tricycle gear in this crabbed attitude. Tricycle landing gear is mandatory since it is dynamically stable and will tend to run out along the runway as shown later in Figure 6-11. However, our gear must be designed to withstand crabbed landing loads, which are really side loads, exceeding those required for normal crosswind conditions. Suitable load criteria are noted in the landing gear comments of Chapter 6.

But what if a sideslip is needed to capture the end of a short runway? This maneuver cannot be performed with interconnected aileron and rudder controls, and was a major source of pilot discomfort with the Ercoupe, one of the original two-control aircraft designs of the late 1930s. Since then, the demands of advanced sailplane performance have provided us with the answer to our problem: wing-mounted, force-balanced speed brakes.

SPEED BRAKES Sailplane speed brakes are located spanwise about midway between the wing root and the inboard end of the aileron, mounted on the top and bottom surface of the wing near the 40 percent chord position. By having the top brake surface move forward into the airflow as it extends while the bottom brake is moved aft by airload, brake operating forces can be kept very low and well balanced. But the drag effect is considerable, depending upon braking surface span and height (area). The great advantage of this control is that local lift is destroyed as the brakes are deployed, but immediately returns when the brakes are retracted.

This feature provides glide-path control by increasing descent rate when the brakes are extended, but restoring the original sink rate when they are stowed. Of course, any intermediate position varies approach descent accordingly. As a result, it is possible to realize more positive approach control with speed brakes than by varying flap extension or slipping. In fact, lift flaps may be used with speed brakes for very precise low-speed descent control; the flaps providing added lift for low approach speed and the speed brakes controlling descent rate and path to accommodate wind variations or pilot judgment. Fully extending the speed brakes following ground contact could apply wheel braking, similar to sailplane operation.

Speed brakes as used here might better be called *spoilers* or *air brakes* since they act to locally destroy lift and so increase rate of descent. But they can also be used to control maximum airspeed if an airplane is caught in extreme turbulence or gusts, or any similar weather situation in which it is a matter of safety to dive away steeply without exceeding maximum airspeed limitations. Therefore, we have the name *speed brakes* or, more properly, *speed control brakes*.

At the present time it is not feasible to adopt these speed brakes, or even a trailing edge spoiler version, for lateral control. Unfortunately, all variable protruding surfaces tested to date for roll control develop a nonresponsive dead spot—and may even exhibit control reversal—at small deflections frequently required for fine correction during instrument flight or landing flare. So our "safe airplane" wing will have sealed ailerons, high-lift slotted flaps, and speed brakes—but no spoiler control system. We will also incorporate 2° of aerodynamic or geometric twist to assure positive lateral control at maximum angle of attack.

WING TWIST This amount of aerodynamic twist may be obtained by selecting a tip airfoil which stalls at 2° higher angle of attack than the root section, and then laying out the intermediate wing sections to blend these root and tip airfoils accordingly. Geometric twist to provide the same result would require twisting the tip chord line trailing edge up 2° relative to the root chord line (known as wing *washout*), and then varying the incidence of intermediate ribs according to their spanwise location. For example, with 2° of washout, a wing of constant airfoil section would have the midpanel rib angled 1° trailing edge up with reference to the root chord line. Whichever method is used, the wing tip area will keep working at greater angles of attack than the root section. This particular design detail assures ample roll control during low-speed flight—an important safety feature near the ground or in busy instrument conditions. However, it will also slightly decrease rate of climb because of reduced lift developed at a given angle of attack.

These stall/spin prevention design details along with other related safety features noted in the Chapter 6 summary have been combined into the configuration of Figure 6-18 presented in that chapter. This side view shows that trim changes with power will be minimized by flow over the horizontal tail, while flap drag forces have practically no vertical offset from the design c.g. location to cause longitudinal trim changes with flap deflection. So even though the thrust line is offset above the c.g., this design will experience minimum trim change with variations in power or flap setting.

REFERENCES 3.1 *General Aviation Stall/Spin Accidents 1967–1969.* National Transportation Safety Board Report NTSB-AAS-72-8, September 1972.

3.2 David B. Thurston, *Design for Flying.* McGraw-Hill Book Company, New York, 1978, chap. 3, pp. 32–33, chap. 7.

4 FLIGHT INSTRUMENTS AND EQUIPMENT

This chapter might be entitled "Collision Avoidance," since we will mainly discuss ways to prevent such accidents. As indicated by the percentages in Figure 1-3, just about every type of accident, regardless of what caused the initial trouble, ends up with the airplane striking something—all too frequently the hard earth. Figure 1-2 lists failure to obtain or maintain flying speed as the largest single pilot error problem, closely followed by failure to see or avoid other aircraft, objects, and obstructions. In fact, these two causes combine to exceed all pilot error powerplant problems. In view of this, let us first consider possible solutions to maintaining adequate flight speed.

Small aircraft usually signal when they are ready to fly; that is, they do if the pilot is familiar with the airplane, the runway is over 3000 ft long, or the day is not hot and the field is not high. Unfortunately, all these favorable conditions may not apply during takeoff, and occasionally none of them will.

The result may produce an accident whenever a pilot tries to pull a reluctant airplane off the ground before it is ready to let go. And the nearer the end of the field appears, the greater is the temptation to become airborne. So we find aircraft staggering into trees or other obstructions at the end of a runway when they could have been cleared *if* the airplane had been allowed to attain flight speed prior to applying so much up elevator.

DENSITY ALTITUDE COMPUTER

Density altitude obviously is a large factor in accidents of this type as is overloading or loading outside c.g. envelope limits. Unless we develop new aircraft configurations such as that shown by Figure 6-18, there is really no way to combat improper loading except through education and application of common sense. But we can reduce the frequency of density altitude accidents by using the density altitude (Denalt) computer to estimate required runway length at local temperature and altitude conditions, and then combine this preflight preparation with a takeoff speed indicator.

For pilot convenience the effects of temperature and altitude upon takeoff distance and climb rate have been put into a simple circular calculator named the *Denalt performance computer*. This name is a contraction of "density altitude," with separate Denalts available for fixed-pitch and for variable-pitch propellers. Both may be obtained from the Superintendent of Documents, U.S. Government Printing Office, Washington, D.C. 20402.

TAKEOFF SPEED INDICATOR

The takeoff speed, or flyaway, indicator would sense required dynamic pressure for safe flight, thereby automatically integrating density altitude and desired takeoff speed for the airplane. Takeoff speed should be higher than stalling speed to provide a safe margin of control and to allow for possible gust conditions, so the familiar stall warning indicator will not supply the signal required since it operates as speed decreases and cannot indicate speeds below stall or steadily increasing speeds.

To simplify use, the flyaway indicator should be developed for each model airplane, including a simple input adjustment compensating for takeoff weight variations up to design gross. Since about 25 percent of all stall/spin accidents occur during takeoff and mainly from attempting flight prior to reaching flying speed, the requirement for adequate takeoff speed indication is apparently very real, while the opportunities for reducing stall/spin accidents and saving lives are correspondingly large.

The principle of the flyaway indicator is basically that of sensing dynamic pressure for the mass flow of air required to safely lift off. This formula becomes:

$$q_{t.o.} = \frac{K\rho V^2}{2}$$

where $q_{t.o.}$ = dynamic pressure desired at takeoff

K = safety factor to provide the desired margin between airplane takeoff speed and stalling speed

ρ = air density (affected by altitude and temperature)

V = airplane velocity, feet per second

In application, a pickup would sense changing values of $q_{t.o.}$ and signal when the required minimum preset dynamic pressure level is reached. This indication could be as simple as a green light on the panel or might be an "up arrow" projected onto a head-up display screen. Either way, takeoff rotation could begin when the signal appeared. By compensating for actual takeoff weight rather than designing the flyaway indicator for maximum gross weight conditions, it would be possible to take advantage of reduced weight to improve short-field performance. This means that to be as useful as possible the flyaway indicator must be designed to be simply adjustable for variations in aircraft weight.

In addition, either separate models of the flyaway indicator or some basic speed adjustment feature will be necessary to compensate for the different takeoff speeds of various aircraft types. For example, one model would be used with a Piper Warrior II, with a somewhat different sensing unit probably required for a Cessna 210. But the basic principle, system design, and readout indication would be identical for all flyaway indicators. So if ever required, field service would be the same for all units. Unfortunately, this instrument is not presently available, but should be in the near future.

AOA INDICATOR

Although the AOA indicator cannot signal lift-off speed, once a plane is airborne it can immediately provide the correct trim angle for maximum rate of climb. With the indicator usually mounted above the instrument panel within the field of pilot vision, the AOA senses proper aircraft trim for optimum operation during climb, cruise, and approach speed phases of flight. If installed in more aircraft, this neglected instrument would certainly reduce the number of takeoff and approach accidents resulting from slow flight, while pointing the way toward efficient cruising trim and accompanying fuel economy. Even though this instrument has been operational since 1918, for some reason it has never received deserved recognition or acceptance (Ref. 4.1).

For AOA operation an external vane or tab senses airflow angle and through electronic conversion reports effective angle of attack for the wing. This information is presented via a readout pointer calibrated to indicate a slow to fast approach speed range, correct cruise trim angle (speed) for airplane weight, and optimum trim attitudes (speeds) for best angle or best ROC. In addition, variations in pointer movement from a desired takeoff climb or approach trim setting will warn of wind shear, gust action, or pilot technique requiring urgent attention. Whatever the cause, corrective action can be taken immediately to preserve the desired flight trim attitude and so preclude an unexpected stall. Safe Flight Instruments of White Plains, New York, manufactures the AOA indicator most commonly used on small single-engine aircraft and on commercial carriers as well, combining the above features with their well-known pre-stall warning indicator. Teledyne Avionics of Charlottesville, Virginia, among others, also produces AOA indicators for small twins and airline transports.

Because of the obvious safety advantages of the AOA indicator in virtually eliminating low-speed attributable to pilot error, as well as providing economical cruise trim indication, we shall include this instrument in our "safe airplane" design specifications. Of course, for aircraft capable of stalling, the primary purpose of AOA is the reduction of stall/spin accidents terminating in collisions with the ground. And if in the process the AOA indicator improves cruising efficiency as well as simplifying final approach technique in VFR and IFR conditions, so much the better. Check for yourself the safety benefits possible from AOA indication. I'm sure pilots frequently flying in gusty or turbulent conditions usually found near the Alleghenys or other mountain chains, or those constantly exposed to IFR operation will find the advantages offered to be very cheap at the current AOA indicator price of about $600 installed. Further and more detailed AOA system operational features will be found in Ref. 4.2 for readers desiring additional information.

HUD DISPLAY   Recent studies by the NTSB as documented in Ref. 4.3 show quite conclusively that instrument approach accidents usually occur after the pilot has seen either the ground, the airport, or the runway area. Once the runway is spotted, the transition from flight on instruments to flight by visual cues introduces the problem of properly returning to instrument guidance if the approach path suddenly becomes obscured when the plane is near or below decision height or minimum descent altitude. Apparently, a sudden return to instrument flight may be accompanied by an unstable approach in which the airplane occasionally contacts the earth or some other obstacle near the flight path, usually with disastrous results.

Since commercial transport and business aircraft normally have two pilots aboard, the recommended procedure for their approaches during severe instrument conditions is to have the pilot flying the airplane maintain visual contact once established, while the second pilot reads and calls out altitude and position along the instrument landing path (high, low, left, right, speed, etc.). If this method is followed, pilot transitions between the distant view outside and the close instrument panel inside are not necessary, reducing the likelihood of vertigo as well as control instability during the period in which eyes are being refocused.

The advantages of a trained crew following this approach system are obvious and should be quite comforting to an airline passenger riding along in the back during solid instrument weather. But what of the solo pilot who, once having sighted the runway, must return to the panel during the final few hundred feet because of sudden fog or a dense rain shower across the approach path? With the recent increase in instrument ratings, safety during final approach becomes more critical and important. Some effort should be directed toward development of reasonably priced instruments or displays capable of reducing pilot workload during the approach while also improving position accuracy along the final glidepath.

While a coupled 3-axis autopilot can do this if properly adjusted, the original price and maintenance costs are prohibitive for the average pilot. A better and more practical design approach would be the development of an instrument display system equally useful for both VFR and IFR operation. This brings us face to face with the HUD panel briefly mentioned in Chapter 3. As shown in Figure 4-1, the basic instrument would have airspeed, altitude, horizon,

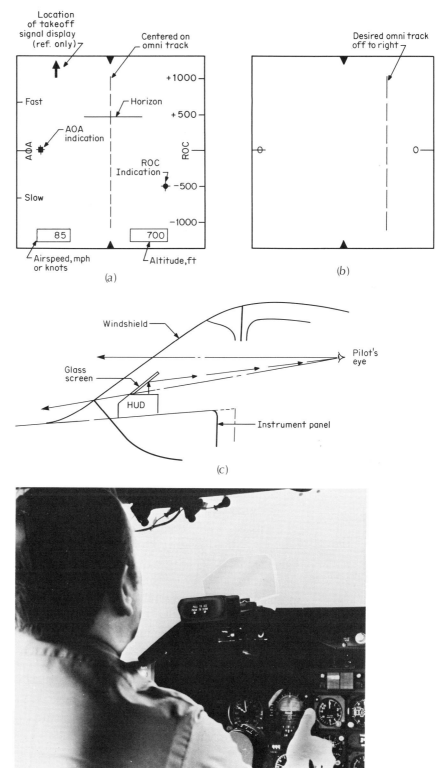

FIGURE 4-1
(*a*) Proposed head-up display (HUD) panel. Data aligned with the pilot's field of vision. (*b*) Display when selected omni track lies to the right (other data omitted for clarity). (*c*) Using proven gunsight technology, data projected onto the clear glass screen can be read by a pilot sighting on a distant runway. (*d*) Sundstrand HUD display screen under flight evaluation.

Sundstrand Data Control, Inc.

AOA indication, ROC, and omni (omnidirectional radio range station) centering projected onto an inclined clear glass screen. The takeoff, or flyaway, signal could also be included, possibly by a green arrow as indicated in Figure 4-1(a). This display is mounted above the instrument panel, centered on the line of pilot vision along a normal descent approach path, a location also useful for takeoff climb data presentation.

The proposed HUD display of Figure 4-1(a) shows a nonprecision approach well centered on the localizer at 85 mph, descending at 500 ft/min through 700-ft altitude. Note that the takeoff signal is shown only to indicate its position on the display screen. It seems logical to also include the omni reading in a form similar to the dashed lines used for road marking. This is recommended because these lines are familiar to everyone and easily read during night driving; they are frequently the only distinguishable guide on a fogbound secondary road. In fact, the simple idea of a bucket of paint and a paintbrush really made night driving practical and every driver knows how much the center and side stripes are missed at night if worn off or covered over.

Since cruising or descending in the fog is similar to night driving in adverse weather, the use of dashed lines for omni tracking should permit more precise control than an inverted needle wobbling between dots. In addition, when the dashed line moves to the right as shown in Figure 4-1(b), we all instantly associate this with being left of the desired path. Conversely, if the dashed line moves left, we immediately realize the aiplane is to the right of the desired track. I suspect movement of the dashed-track line to the right will generate the more rapid corrective response, since this gives the sensation of driving on the wrong side of the road.

The use of dashed-line tracking on the HUD should also be useful for cruising because it permits scanning of the flight area while monitoring the omni course, and doing so without dropping vision down to the instrument panel when running in and out of clouds or flying at night. Of course, engine instruments would require occasional panel scanning, but this could be done when convenient.

Having AOA indication right along the takeoff climb or approach line of sight permits rapid and safe attitude control correction, supported by ROC and airspeed values. This instrument arrangement converts the need to scan many instruments into one centralized readout area that may be peripherally monitored while concentrating on the approach. If desired, glide slope could be added as a dashed horizontal line moving up or down from center to indicate aircraft vertical position relative to the precision approach path.

For full sophistication, directional gyro (DG) or horizontal situation indicator (HSI) readout could be projected onto the screen, but this might create congestion that would defeat the basic purpose of our proposed instrument; it is primarily intended to supply necessary data for safe climb and approach control under all flight conditions from VFR through IFR. For example, having the AOA presentation directly in the field of vision supplies instant indication of ambient wind shear and gust activity or any other deviation from the optimum flight path, providing a wide margin of safety above low-speed stall/spin or mushing-into-the-ground-type pilot error accidents.

There could be some debate concerning the desirability of presenting altitude readout on the HUD panel since this data normally requires encoder or similar signal from the altimeter, whether it is the one supplied with the airplane or an altimeter portion of the HUD instrument. Aircraft operating in TCAs will have altimeter encoding output available; however, most pilots do not have or require equipment of this caliber. It should be possible to present altitude readout directly from static pressure without costly electronic transducer data conversion, similar to the way airspeed readings will be obtained. And it certainly would be advantageous to have altitude displayed within line of sight when searching for the end of a foggy runway.

While this type of HUD instrument is not presently available for small general aviation aircraft, experiments have been conducted in Europe leading to glidepath control through an approach slope alignment instrument worked in association with the throttle for altitude correction. Known as *Spotland,* this system has been flown successfully (Ref. 4.4), although it is not ready for market at this time. The proposed Spotland system price is about $4000, which seems high for the average general aviation owner and about 4 times the target price of the basic HUD reviewed above.

Gunsights used on fighters since World War II are based upon the reflective principle of Figure 4-1(c), so the technology for development of a simple HUD instrument has been around for some time. This background can now be directed toward a basic flight instrument that would simplify flight procedures while materially reducing pilot error–induced takeoff climb and approach accidents. With some effort it could be available during the 1980s. One such test flight evaluation display is shown in Figure 4-1(d).

AUTOPILOTS Further simplification and reduction of pilot workload is possible through use of an autopilot—a form of selective copilot assistance. Although a smoothly operating 3-axis autopilot would be the ultimate control system, what is basically required is someone or something to hold the wings level when a pilot is trying to work out a revised estimated time of arrival (ETA), or find alternate approach plates, etc., in IFR conditions.

Similar support would also be helpful in VFR conditions when a pilot needs to check charts to spot cross-country position, look for something in the cabin, or just rest on long trips. In other words, what is needed is some assistance for 1 or 2 min in bad weather as well as a relaxing aid in VFR conditions.

The so-called wings leveler will do this handily and has been standard equipment on Mooney aircraft for some time. Usually consisting of a single gyro spinning about a transverse axis (spanwise axis of rotation), this basic autopilot is capable of keeping the wings level in moderate turbulence as well as smooth air conditions. That is, it will if it is properly matched to the airframe stiffness and aileron response characteristics. I have personally experienced an improperly coupled single-axis autopilot that within 30 sec took the airplane I was flying from level cruise at 120 mph into a steep spiral dive exceeding 150 mph. When encountering snowstorm turbulence along the IFR route, I had decided to find out how well the autopilot was working before I might really want it during final approach. So I turned the system on amid mild turbulence and watched in fascination as the autopilot attempted to bring the wings level by imposing corrective aileron displacement almost in phase with the turbulence. In other words, aileron action was applied in the same direction and frequency as the upsetting gusts. As a result the control corrections became increasingly violent; being in phase with the turbulence, they kept adding to lateral displacement rather than correcting the condition. At 150 mph I disconnected the autopilot and took over manually to climb back to assigned cruise altitude.

The moral of this story is that an autopilot should be checked when it is not needed but can be carefully observed for proper operation. Clear weather is the time to find any defects, not during final approach when attention to other items might let the autopilot set up a terminal spiral from low altitude. So, while dependence upon a properly functioning autopilot can be of great assistance during long cruising legs or on a dusty final approach, trust only a recently checked system.

Readers familiar with gyros will realize that a mass spinning about a spanwise horizontal axis is as resistant to directional (yawing) motion as it is to lateral displacement. This means the single-axis autopilot can also be used for directional control if the added sophistication and control complexity, including cost, are acceptable. While heading can be maintained reasonably well with just a wings leveler correction based upon compass input or omni-bearing selection, the output from a 2-axis (lateral and directional) autopilot will provide better and more precise course control.

Throwing in another gyro, rotating about a vertical axis, will permit pitch control, a very nice feature for long legs in any kind of weather, but particularly useful for setting up controlled rates of climb or descent as well as holding assigned altitudes in severe IFR conditions. Obviously, a gyro rotating about a vertical axis has resistance to any displacement from the horizontal. This means a single, vertical-axis gyro could be used for both pitch and roll control, although heading hold might not be as precise as from a gyro mounted to spin about a spanwise axis.

As a result of all these possible gyro combinations, a number of manufacturers provide autopilots with varying degrees of complexity, usefulness, and cost. Realizing that most small aircraft require only simple assistance to keep the wings level, some dedicated lightplane enthusiasts at NASA Langley Research Center have been working to perfect the development of a poor man's autopilot. Based upon a simple fluidics design completely free of gyros of any type, this system is shown in Figure 4-2.

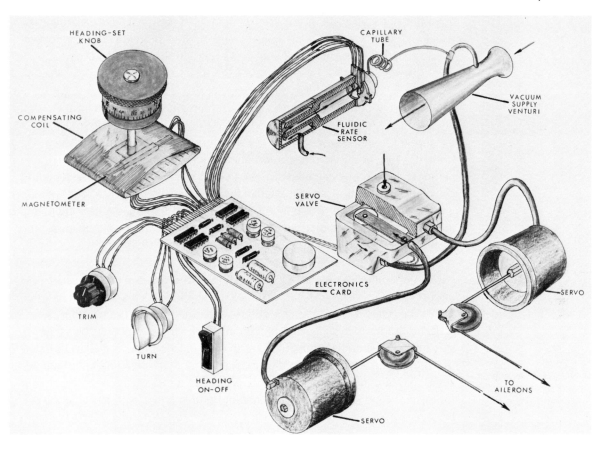

HEADING-SET
KNOB

COMPENSATING
COIL

MAGNETOMETER

TRIM

TURN

HEADING
ON-OFF

ELECTRONICS
CARD

SERVO
VALVE

CAPILLARY
TUBE

FLUIDIC
RATE
SENSOR

VACUUM
SUPPLY
VENTURI

SERVO

TO
AILERONS

SERVO

FIGURE 4-2

Electrofluidic wing leveler and heading reference flight tested at Langley in 1973.

NASA Langley Diagram

According to H. Douglas Garner, developer of the electrofluidic wing leveler portion of this simplified autopilot, the field of fluidics is concerned with the flow of fluids in aerodynamically shaped ducts and chambers to perform functions such as sensing, signal amplification and processing, logic manipulation, and mechanical actuation normally performed by electronic and electromechanical devices. As a result, properly functioning fluidic controls would be cheaper and more reliable than the complex units they can replace, but getting a fluidic autopilot to operate as required has been a basic problem. This goal is now in sight (Ref. 4.5), so we should expect to have cheap and reliable 2-axis autopilots available for light aircraft use during the 1980s. Freedom from routine flight tasks provided by this level of equipment will make flying much more pleasant, while flight safety will also be improved through reduction of pilot error accidents caused by improper handling in bad weather.

If we stop a moment to reconsider, all four items just reviewed relate directly or indirectly to weather and its side effects. For example: safety benefits offered by the takeoff speed indicator increase as density altitude conditions become more demanding; AOA trim speed and attitude readings are particularly helpful during flight through gusty or turbulent air; while the HUD indicator and autopilot put it all together for safer operation in marginal VFR or critical IFR weather. Since weather is a timeless and tireless problem, this equipment would help to reduce its adverse effects upon all phases of flight.

OTHER INSTRUMENTS Another new, although rather expensive, instrument designed to combat weather is the Ryan *Stormscope* — a form of electrical energy receiver intended to pinpoint thunderstorm activity. By selecting storm detection ranges varying from 40, 100, or 200 nautical miles ahead along the flight path, a pilot sitting behind a Stormscope can take evasive action to safely fly around thunderstorm cells. This information is pinpointed in location and intensity by a series of little green dots spotted on a display tube. Since this display has a grid indicating thunderstorm range and direction in relation to the airplane heading, the Stormscope clearly shows the course away from electrical activity.

Whether this information is worth the price of around $6500 installed will depend upon the frequency of IFR operation in thunderstorm areas, or just how important equipment cost and time are in relation to completing scheduled flights regardless of weather conditions—completing them safely, that is.

If the Stormscope or similar pickup and display system could be carried a step further, it should be possible to scan for local air traffic as well as thunderstorm location. A single, simplified display capable of providing both thunderstorm and aircraft proximity guidance would be a most valuable safety device for business and small feeder-line aircraft frequently engaged in marginal VFR or severe IFR operation. The British Civil Aviation Authority (CAA) has analyzed the frequency of midair collisions over the 30-year period from 1946 through 1976 (Ref. 4.6), finding that collision risk has remained nearly constant during this time although the number of fatalities has about doubled. Pilot error is credited with 56 percent of these accidents and controllers with 13 percent, with the remaining 31 percent assigned to other causes. As "inadequate lookout" was found to be the most common contributory pilot error, it is obvious that we are long overdue for some better or more reliable form of spotting nearby aircraft in order to supplement the view from a rather blind cabin.

Everyone is familiar with the helpful location of aircraft within radar control areas communicated to us by approach control personnel; and even if we can't find the other airplane much of the time, the challenge does keep our adrenaline active. However, pilots frequenting congested areas or skirting major TCAs really need more positive and rapid indication of local traffic; thus some aircraft-proximity-indicating instrument similar to the Stormscope should be developed for their use and safety. We should look for this protection during the 1980s.

In the meanwhile, the best midair collision prevention for most of us will remain a swiveling neck, preferably coupled with considerably improved cabin visibility, as per the design shown in Figure 6-18 or some similar configuration. And the 180° turn away from apparent thunderstorm activity still remains the best basic protection when faced with unpredicted weather.

With the current comparatively high level of aircraft engine and propeller reliability, which translates into operational safety, future small aircraft development as outlined in Chapter 6 will not only incorporate powerplant features of the next chapter but also benefit to a considerable degree from advanced flight instruments and systems outlined in this chapter. In fact, it is expected that advanced aircraft and flight systems development will move forward together during the 1980s to provide more easily handled, more efficient, and safer aircraft. While certain component items will cost more, increased production should serve to keep final costs at today's levels (in comparative dollars, of course).

REFERENCES

4.1 Fred H. Colvin, *Aircraft Mechanics Handbook.* McGraw-Hill Book Company, New York, 1918.

4.2 Dan Manningham, Flying Angle of Attack, *Business and Commercial Aviation Magazine,* September 1975, p. 54.

4.3 *Flightcrew Coordination Procedures in Air Carrier Instrument Landing System Approach Accidents.* National Transportation Safety Board Report NTSB-AAS-76-5, August 1976.

4.4 Barry Schiff, Poor Man's Autothrottle, *The AOPA Pilot,* January 1978, p. 55.

4.5 (a) H. Douglas Garner and Harold E. Poole, *Development and Flight Tests of a Gyro-Less Wing Leveler and Directional Autopilot.* NASA TN D-7460, April 1974.

    (b) H. Douglas Garner, *Applications of Fluidics to Light Aircraft Instrumentation and Control.* Society of Automotive Engineers Report 740351, April 1974.

    (c) Don Hewes, Development of a Poor Man's VFR Autopilot, *Sport Aviation,* May 1978, p. 24.

    (d) H. Douglas Garner, Construction Notes on Electro-Fluidic Wing Levelers, *Sport Aviation,* June 1978, p. 32.

4.6 *Analysis of Mid-air Collisions.* Civil Aviation Authority Publications Paper 76041, 1977. Address: CAA, Greville House, 37 Gratton Road, Cheltenham, Gloucester, England. Price, $5 in 1978.

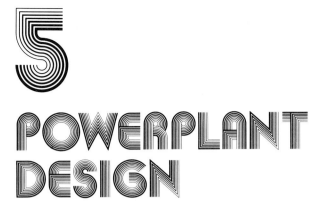

5

POWERPLANT DESIGN

WHAT CAUSES POWERPLANT FAILURES?

As would be expected, if aircraft engines stop it is usually during flight—and the only way to go is down. Unless accompanied by fire or some sort of explosion throwing shrapnel around, powerplant failure is, mechanically speaking, a relatively harmless event, but a hair-raising experience for those in the cabin. Without doubt, all pilots automatically and instantly associate engine failure with a severe crash. Fortunately, this is not normally the case.

FUEL MISMANAGEMENT

While only about one-tenth of all aircraft experiencing powerplant failure are involved in fatal accidents, even this rate is unacceptably high. Considerable improvement is possible since *fuel mismanagement* is by far the greatest single cause of powerplant failures. Fuel mismanagement consists of such pilot error factors as switching to empty tanks, turning the selector valve to "fuel off" instead of to a different tank, landing while drawing fuel from the wrong tank (not following placard instructions), taking off without adequate fuel, and running out of fuel due to inattention or bad weather.

If we design dual tank systems that properly feed as a single tank and then use a selector valve with "left," "both," "right," and "off" positions that could be placed on "both" for all normal flight conditions, the pilot error potential for switching to the wrong tank or ever selecting the "fuel off" position would be virtually eliminated.

The few records of engines quitting while on the ground usually result from taking off with the fuel selector valve in the "off" position. The amount of fuel in a carburetor bowl can be sufficient to permit starting the engine, checking the magnetos (mags) and heat, and beginning the takeoff run—before dead silence occurs. Again we are faced with a classic pilot error—neglecting to use the preflight checklist. In preference to tangling with trees or poles at the end of the runway, this sort of engine failure usually results in injured pilot ego plus a ground loop with some minor airframe damage.

But if a takeoff is started on a virtually dry tank, the powerplant may stop operating at 50 to 100 ft when right over the airport boundary (with trees, poles, or houses below). This can be more serious than engine failure at cruising levels, since there may be no suitable landing site available from such a low altitude. And there is not sufficient height to attempt a turn back to the airport, although many pilots in this situation try to return home by stretching their glide—ending their takeoff in a spin caused by engine failure.

Fuel mismanagement can also include improper distribution of fuel, resulting in an aft or forward c.g. located outside the approved weight and balance envelope. This condition may result in loss of control during takeoff or climb; during approach, it could result in an untrimmable or uncontrollable airplane, particularly when the flaps are lowered with a forward c.g. This is another type of accident in the pilot error category. Why not design aircraft so the approved c.g. balance limits can never be exceeded regardless of how much fuel is taken aboard, or in which tanks?

Of course, flying overweight is a prime pilot responsibility, which must include consideration for density altitude conditions, fuel weight for the distance to be traveled, passenger weight, and baggage load. If too much fuel is loaded into oversize tanks, the airplane may not become airborne, and the resulting accident is not the fault of the designer or the airplane.

Pilot mismanagement of fuel can range from too much aboard to too little delivered to the engine, known as *fuel starvation*. Fuel starvation may result from switching to an empty tank or to the ''off'' position; so look at the handle of the selector valve and positively read where it is pointing when changing tanks.

It is also possible to experience fuel starvation from factors other than pilot error. For example, 80- and 100-octane fuel tanks have been filled with jet fuel, which does not burn in existing reciprocating engines. This is considered a service personnel–induced accident rather than pilot error, but it is still just as severe a problem when incurred. Why not index fuel nozzle size and shape with tank filler openings for different types of fuel? After all, the value of any airplane justifies replacement of $50 worth of tank filler caps, particularly when flight safety is involved.

UNEXPLAINED POWER
FAILURE

Fuel starvation can occur simply because fuel doesn't get to the engine. This failure may be due to a slug of water frozen in the delivery line at altitude or result from tank debris clogging strainer screens and injection systems. Probably a number of the *unexplained powerplant failure accidents,* the second largest category after fuel mismanagement, result from fuel line or induction system icing that melted prior to inspection of the accident. I also suspect that some unsolved accidents were caused by pilots who would not admit to improper powerplant operation and control. While we can do little about judging pilot confessions, prior to flight we can drain the tanks to remove water. Also, designers can place fuel strainers where they may be readily removed for cleaning, and provide inspection openings permitting fuel tanks to be easily cleaned at regular intervals.

MPROPER OPERATIONAL PROCEDURES

Going further into classified powerplant accidents, we find the third largest cause attributed to *improper operation of the powerplant and powerplant controls.* This category does not include previously discussed mismanagement of fuel, but, for example, would cover selecting mixture idle cutoff instead of some desired propeller pitch or power change, misapplication of carburetor heat, neglecting to follow the landing checklist requirement of full rich mixture when descending from cruising flight (and so conking out on final), incorrect operation of turbocharger controls, and similar items.

Improper operation of the mixture control is responsible for many powerplant accidents of this type. With today's volume of general aviation engine production, it would seem quite logical and relatively inexpensive to provide automatic mixture control for all carburetor engines. While mixture cutoff could still be used for shutdown with "full rich" selected prior to starting the engine, all carburetor altitude and power conditions would be monitored and controlled automatically. The automatic mixture feature should provide optimum fuel economy, relieve the pilot of another burdensome problem, and remove one more opportunity for pilot error.

Induction systems can become clogged with ice or other materials, shutting off the air supply and causing engine failure. Severe icing conditions can coat air filters in a very few seconds, immediately choking the engine to death. Such failure can be an equally serious problem with normally aspirated (unblown, sea-level engines), fuel-injected, and turbocharged engines since they all require air for operation. Powerplant designers can do much to relieve this problem by using flush air intakes, locating carburetor air filters in an area surrounded by heated air, and providing automatically opening alternate air doors downstream from the inlet—again removing a potential source of emergency action that might be mishandled and result in another pilot error powerplant accident.

A recent NTSB study covering a 5-year operating period found that 360 accidents were attributable to carburetor icing either as the cause or as a contributing factor. Some aircraft induction systems are more subject to icing than others, and these may require application of carburetor heat any time the power is reduced to near idle speed regardless of outside weather or temperature.

The remaining powerplant failure accidents consist of many different items of comparatively small percentage each. The largest of these is the category of engine structural failures, such as: crankshafts, connecting rods, gears, magnetos, cases, and valves; turbocharger and impeller blowers; carburetor and fuel system components; throttle controls; and propeller blades.

Since we have design control over all these items, it is possible to reduce the number of accidents attributed to them. For example, propeller blade failures can be virtually eliminated by testing for engine-propeller compatibility prior to final selection of these major components for any new airplane. It is a relatively simple matter to determine that no high blade stresses occur over the entire propeller rotational speed spectrum, as we shall see.

The requirement prohibiting engine operation at certain rpm (revolutions per minute) values can be eliminated by properly damping the engine internally or by modifying the propeller blades—or through both approaches; yet many aircraft fly about with red bands on the tachometer dial restricting engine operating speeds. It seems strange that aircraft can be certified with power restrictions which increase cockpit workload during tight going, or which, if ignored, may result in a blade failure blamed on the pilot for operating at a prohibited rpm (probably at a time when there was so much to do that the sound of engine operation was welcome at any rotational speed).

Oil cooler and oil line failures are relatively infrequent, thanks to modern fatigue testing procedures and fireproof materials. But fuel vents, drains, and tank caps need further study and improvement. Exhaust system manifolds and mufflers regularly cause a few accidents, and should probably be subject to more rigorous vibration testing prior to acceptance for use on any new design.

As you can see, any part of the powerplant may cause trouble from time to time. And the hand of fate is not particular about the number of engines per plane. Powerplant failures occur with similar frequency in both single- and twin-engine aircraft, although the percentage for twins is slightly worse, based upon percentages of total aircraft (Ref. 5.1). Further, the final results are somewhat more damaging for the twins since loss of control could cause a powered descent with one engine operating to wind into a severe spin condition.

Many fatal accidents involve fuel fires after a crash, so every effort should be made to keep fuel storage and plumbing away from the cabin as well as to provide a positive means for switching off the electrical system at impact. Both of these requirements can be satisfied through proper design.

POWERPLANT PROBLEMS AND IMPROVEMENTS

APPROVED ENGINES AND PROPELLERS

The use of newly certified engines on production aircraft has always been a source of difficulty, recently highlighted by Cessna's trouble with the redesigned 160-hp Lycoming installed on the model 172 Skyhawk. The resulting modifications seemed less safe than the 80-octane fuel shortage problems this engine was supposed to correct. Fortunately, there are many well-proved engine models in the 115- to 300-hp range, so the use of an established and accepted engine is the number one powerplant safety recommendation.

This is closely followed by the need to select engine and propeller combinations free of any placarded rpm restrictions throughout the entire operating

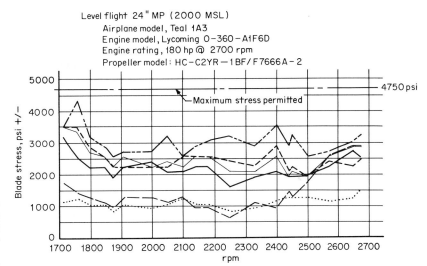

Level flight 24" MP (2000 MSL)
Airplane model, Teal 1A3
Engine model, Lycoming O−360−A1F6D
Engine rating, 180 hp @ 2700 rpm
Propeller model: HC−C2YR−1BF/F7666A−2

FIGURE 5-1
A propeller blade survey indicating acceptable stress levels; rpm placarding would not be required for this flight condition.

Hartzell Propeller, Inc.

range. Engine and propeller manufacturers are willing to suggest compatible engine and propeller models offering unrestricted operation, and will also work with designers to assure maximum delivered thrust for optimum climb and cruise performance. In order to demonstrate that a powerplant is free of restricting vibratory stresses, one propeller manufacturer insists upon instrumented flight testing of every new airplane design or powerplant modification using their products. The safety advantages of such investigations have been well proved by fewer blade and crankshaft failures; in fact, the FAA will not grant powerplant design approval for new aircraft unless instrumented propeller/engine flight test data indicate acceptable propeller blade stress levels.

Figure 5-1 is an actual flight record of favorable test data obtained during a recent certification program. The various curves represent stress readings taken at different points along the propeller blade, all of which must remain below the maximum allowable stress level throughout a series of different operational conditions. This data represents just one of the series of successful tests.

Of course, all this is time-consuming and costly, but it is necessary to provide powerplant operational safety. For some reason, placarded (prohibited) rpm red-band ranges always seem to fall in the rpm range selected for IFR approach speed and rate of descent. The elimination of restricted rpm placarding is one less item a pilot must monitor during a "tight" approach—as well as one less type of powerplant failure attributable to pilot mismanagement (from operation at prohibited rpm).

AUTOMATIC MIXTURE CONTROL The use of altitude-compensated carburetors seems long overdue and urgently required to correct situations in which descent from altitude is continued into final approach with the mixture still lean, usually ending in engine failure from fuel starvation. Since opportunities for this type of pilot error would be eliminated by automatic carburetion there seems little justification for not using pressure carburetors and injectors monitored by automatic mixture controls sensing power and altitude conditions. As a bonus, such devices would provide optimum fuel setting and efficiency throughout the flight profile without depending upon pilot vigilance for proper mixture control.

SINGLE LEVER
POWER CONTROL

Believing that if slight improvement is good, doing the whole job is better, Woodward Governor Company developed the Single Lever Power Control for powerplant operation. Now approved for one model Continental engine used in the Beech Bonanza, and at present limited to fuel-injected engines, this design includes automatic propeller pitch as well as fuel flow and mixture controls.

Shown in Figure 5-2, the Single Lever Power Control is operated by a throttlelike power lever. As described by Woodward, the system has three major components: a combined throttle body and fuel control, a Woodward propeller governor, and mechanical linkages to connect them. By setting the mixture control lever to "automatic" (Figure 5-3) and manually selecting a desired rpm setting with the power control, the system automatically regulates manifold pressure and fuel flow. Whether the aircraft is climbing, cruising, or descending, the control adjusts for changes in ambient conditions and altitude to maintain a constant power setting below engine critical altitude. Above critical altitude, when the throttle is wide open and full power can no longer be maintained, the unit automatically adjusts fuel mixture in proportion to the power available.

This new system is totally mechanical and has no electrical components to burn out or fail. It also features manual mixture control settings for safety

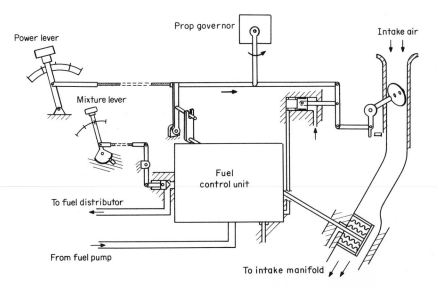

FIGURE 5-2
Single Lever Power Control system schematic diagram. One lever controls all powerplant functions.
Woodward Governor Diagram

FIGURE 5-3
Beech Bonanza with the Woodward Governor single control for powerplant operation.

Woodward Governor Photo

backup and as a fail-safe mode. Figures 5-2 and 5-3 show that the pilot has direct authority over throttle and mixture if a failure occurs in any of the sensor units. This is an approved automatic powerplant control device that can take much of the burden from every pilot's shoulders, and along with it many potential opportunities for pilot error. Versions of this device should be used on all newly approved aircraft and will be included in our "safety airplane."

Since reasons for standardizing all controls and instruments will be discussed at some length in the following chapter, noting the need for powerplant control and indicating instrument standards will serve our purposes here. But this is an important safety item that should be implemented by industry as soon as possible.

ALTERNATE AIR DOOR While ice is to be avoided at all costs by the average pilot, particularly in small aircraft not equipped for such weather, there are instances and times when ice is there; since it is unpredicted and unexpected, being placarded to stay clear of ice is no help if you are caught. One of the more serious results of such encounters can be almost immediate icing of the engine air intake filter. If this condition is recognized in time, the required alternate (hot) air source can be selected along with a slight loss of power. Since even fuel-injected engines have been found to experience induction icing, newer fuel-injected power-plant installations have alternate hot air, although many older designs do not. Any airplane not equipped with manual selection of alternate hot air should be modified to provide this feature as one of the important powerplant safety items.

But conditions in the cabin during a surprise ice attack may well preclude any thought of alternate air selection until the engine becomes silent. To prevent this very real and dangerous situation, an automatically opening alternate air door should be installed in the engine induction system just downstream from the filter as shown by Figure 5-4.

The alternate air door should be lightly spring-loaded to remain normally closed. When the inlet air filter is clear, ram air pressure will also help keep this door shut; when the filter is blocked, engine induction system suction will open the automatic door to obtain air necessary to keep the engine running. This air will be taken from inside the cowling below the engine and so will be warm and free of moisture—a combination unlikely to cause induction system icing. If some engine rpm drop is noted when the automatic door opens, the usual alternate hot air system may be selected. Either way, the engine should continue running safely despite ice buildup on the inlet filter.

Because debris such as mud, wastepaper, birds, etc. can also clog the inlet air filter, the automatic air inlet door is really a safety item needed during all flight conditions from takeoff to landing. As such, it should be included in every modern powerplant design. By automatically taking care of emergency induction air requirements, we can eliminate another cause of pilot error accidents. In such situations the error probably would be blamed upon "improper operation of powerplant controls," but the automatic alternate air inlet door acts as a timely correction. Of course, ice may still be building up on the wings and tail, but with the engine continuing operation there should be some opportunity to select a safer flight level or an acceptable landing area. Without an engine, there is no choice but straight down.

FUEL SELECTOR
OPERATION
For many reasons fuel systems cause problems out of proportion to their basic complexity. Part of the trouble is that some fuel systems are unnecessarily complex, either because of the original system design or from tanks, lines, and

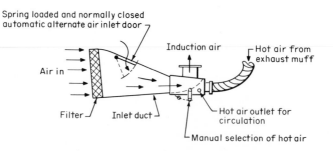

FIGURE 5-4
Position of automatic alternate induction air inlet—carburetor or fuel injection engines.

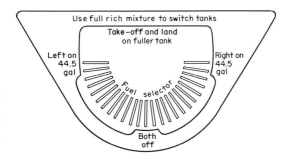

FIGURE 5-5
Fuel selector valve placard for Cessna model 210G. The "both off" position shown here is the "both on" for Cessna models 172 Skyhawk and 177 Cardinal.

valves added over a period of time as the airplane developed—or from both causes. The current Aerostar fuel system problems and related accidents appear most baffling and are probably the result of complex fluid flows developed within this multiple tank system design.

Ideally, a pilot should find the same configuration and location for the fuel selector valve in every single-engine airplane produced, and a similarly consistent arrangement in all twins. Unfortunately, virtually every model airplane has different fuel selector positions, location, and operation. Some models of the same manufacturer even have different selectors for basically similar aircraft—it is small wonder that accidents occur. Just being "checked out" in a model is not sufficient and does not replace experience when an emergency arises.

Let us consider the fuel selector valve used on the Cessna 210. This is a single-engine airplane looking much like Cessna's 172 Skyhawk or 177 Cardinal but with more power and retractable landing gear. As shown by Figure 5-5, the fuel system is *shut off* when the selector valve is centered on the 210 placard. But placing the selector in that position for models 172 and 177 permits *flow from both wing fuel tanks*. Any pilot with considerable 172 or 177 time, but recently checked out in a 210, would be inclined to select the center valve position if the engine coughed when on either tank—thereby turning off the fuel system. Or worse yet, if the word "off" is worn from the 210 fuel selector placard, the remaining word would read "both", and we again wind up with a dead engine as in the crash reported in Ref. 5.2.

FUEL SYSTEM DESIGN   If a wing-tank fuel system carries equal volumes of fuel in left and right tanks, the system should be designed to permit even flow from both tanks. The tanks should be vented to act as a single unit by using a layout similar to Figure 5-6. (This fuel system is not intended for aerobatic aircraft, which require more complex venting to prevent fuel siphoning during maneuvers.) The vent lines must be larger than usually used with small aircraft fuel systems to permit free airflow between tanks; otherwise the tanks may drain unevenly, causing annoying lateral unbalance.

When this fuel system is used, the selector valve will normally be placed on

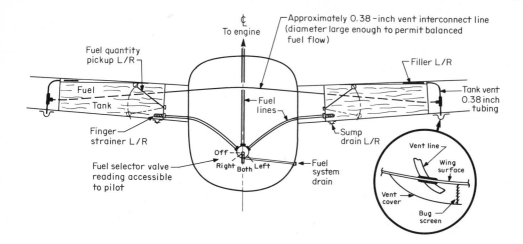

FIGURE 5-6
Fuel system schematic diagram showing vent line interconnection permitting both tanks to act as a single tank.

"both tanks" for all flying. However, in case of tank leakage the selector valve should also permit independent selection of either tank. The resulting fuel selector valve placard should read "left," "both," "right," and "off" in that order with detents located accordingly. With this arrangement, the selector valve cannot be aligned with the "off" position unless the fuel system is purposely shut off. By having both tanks behave as one, the pilot need not be concerned about switching tanks every half hour or hour—providing one less operation which, if overlooked, could become pilot error as a tank runs dry on final.

To prevent filling tanks with the wrong type of fuel, the automobile and petroleum industries agreed to develop hose nozzles and tank filler openings that are compatible (as for unleaded fuel). While this may have been the result of government regulations, it happened nonetheless. Before the FAA or NTSB issues a similar mandate for aircraft, the general aviation industry should establish nozzle and filler opening requirements for the different octane and jet fuels to prevent filling tanks with the wrong fuel. Unfortunately, a number of recent accidents have resulted from jet fuel's being pumped into reciprocating engine tanks. Since reciprocating engines require high-octane fuel that will ignite (explode) in the cylinders, they cannot operate on lower-octane (oil-burner-type) jet fuel. In addition, the possible use of 80-octane fuel in 100- to 130-octane tanks is suspected as a cause for occasional loss or critical reduction of power during takeoff.

Labeling a tank opening is not enough; accidents occur because the human element is involved in a chain of events—and aircraft tanks are frequently fueled when the pilot in command is in a waiting room or on a business visit. But if the wrong nozzle could not be fitted into the filler opening, the wrong fuel could not be pumped aboard. Matched nozzles and aircraft tank filler openings for the various grades of fuel would eliminate a rather constant cause of pilot concern as well as another source of powerplant accidents.

Clogged fuel tank vent lines can result in collapsed and ruptured fuel tanks or ruptured fuel pump diaphragms. We were well into one certification flight test program, in fact in the final stages of fuel system operational tests, when our FAA flight test pilot had to make a dead stick landing from 7000 ft. We soon found that during this climb test the fuel pump was steadily forming a vacuum in the wing leading edge fuel tanks, until finally the pump diaphragm burst. Of course, the engine stopped as soon as the carburetor float chamber ran dry; but since the test was conducted while we were circling our field, an emergency landing was no problem.

The vacuum in the system was so severe that the fuel filler caps could not be released until the vent lines were cleared of mud wasp packings—both tanks having similarly plugged vents. The wing leading edge tank skins were sucked down more than ½ in. in contour between ribs, but popped back into shape when the vents were reopened. Since insect action of this type may not be detectable during routine preflight inspection, a vent cover and bug screen similar to that shown in Figure 5-6 is highly recommended. This is not a gadget, but a safety feature of value in any climate; bugs of some sort always seem to be building mud nests wherever warm weather occurs.

This same vent screen is also useful for lightning strike protection. While the effects of lightning on small aircraft as well as suitable protective design features seem to be nebulous subjects at this point in aviation development, there is positive indication that a fuel tank vent outlet located near the geometric center of a wing panel lower surface offers more protection than one located near or at the wing tip. And screening placed over the vent appears to provide further protection by separating passing lightning from fuel fumes. For readers interested in lightning behavior, one of the few books ever written on this subject is listed in Ref. 5.3. While this is not of recent date, all material and discussion remains pertinent, including a section on aircraft protection.

Every year a consistent percentage of powerplant-related accidents result from extending flight time right up to completely dry tanks. In fact, I know of one pilot who has succeeded in doing this more than once. One good solution for preventing such pilot error accidents might be fuel capacity exceeding human bladder capacity. But more seriously, we obtain a safer IFR airplane and more efficient VFR operation through increased range capability.

To be useful, a personal airplane should have 5 hr endurance at 65 to 70 percent cruise power, permitting 4 hr of flight with 1 hr reserve for IFR alternates, or more distance if pilot, passenger, and weather conditions permit. Seaplanes should have closer to a 7-hr range, if possible, for flying 3 hr into that remote camp and then 3 hr back out—plus 1 hr reserve for the unexpected.

Such fuel capacity does not require an overly large airplane because small engines powering small aircraft have correspondingly small fuel consumption rates. Stepping up the scale a bit to 250 hp, we find about 13.5 gal/hr required to support 65 percent power cruise. That amounts to 67.5 gal for 5 hr operation; or more likely a 70-gal capacity to allow for residual fuel that never seems to be usable. This volume amounts to 420 lb of useful load and probably requires a 2700-lb-gross-weight airplane, which is not overly heavy or large for a 250-hp powerplant. Thus, another important powerplant safety criteria is sufficient fuel capacity to permit practical and useful operation.

Among other dos and don'ts for fuel system design, each tank capacity indicator should be set most carefully to red line zero when 2 gal of usable fuel remain in the tank. Also, sight gauges connected to fuel tank outlets or delivery lines should be avoided if possible because they can rupture or be broken off, followed by complete loss of fuel.

The engine cowling requires careful interior design to eliminate all opportunities for fuel to become trapped in remote pockets or behind reinforcing angles, etc. If fuel collects because of overpriming or flooding, a backfire can start a healthy engine compartment fire—and one that may not be immediately noticed in the cabin. As part of your own safety procedure, be sure all fuel promptly and completely drains overboard from the cowling interior. If in doubt, you can check this simply by splashing water inside the engine compartment and watching it drain away properly—or collect and remain here and there to cause trouble.

DETAIL FEATURES   The engine oil breather line is another source of powerplant problems. Usually ignored as it silently vents the crankcase overboard, the oil breather line can freeze solid by buildup of entrapped moisture from exposure to cold air. When it does, the crankcase or gasket may blow, causing complete loss of engine oil. For this reason, the entire length of the line should remain inside the warm engine compartment, with the breather exit located clear of the airframe but immersed in the flow of cooling air exiting the cowling. Check to be sure yours is properly installed.

To help lower exterior and interior noise levels, and so also reduce pilot fatigue, aircraft must be equipped with exhaust mufflers. Automotive-type mufflers are not satisfactory because an extremely large and heavy unit would be required to reduce exhaust back pressure to FAA permitted limits of 1 psi maximum (see Ref. 5.4). Because automobiles normally operate at a small percentage of their total power, automotive muffler back pressure is acceptable for automobiles; but aircraft engines seldom operate for extended periods of time below 50 percent power, and of course routinely use 65 to 75 percent cruise power.

This means an automotive muffler rated for 200 hp will develop high back pressure on a 200-hp aircraft engine, rapidly developing valve and other accident-causing problems. In fact, a 400- to 450-hp automotive truck muffler would be required to provide acceptably low exhaust back pressure for a 200-hp aircraft engine—obviously an impractical size and weight for small aircraft.

In view of this, considerable research has been undertaken to develop special muffler designs for aircraft use. Because most cabin and engine induction air systems are heated by passing outside air over the exhaust stacks, the better

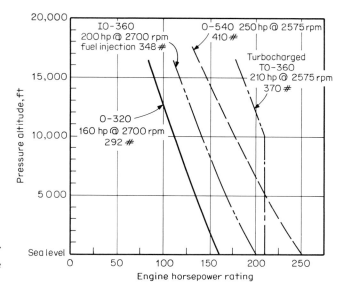

FIGURE 5-7
Power variation with altitude for some production engines.

muffler systems are made of stainless steel in order to reduce the rate of exhaust stack oxidation, and so prevent carbon monoxide from escaping and mixing with heated cabin air. These muffler/exhaust systems are quite effective and are used on most modern production aircraft, including our "safe airplane" design.

SUPERCHARGERS   Exhaust-driven turbosuperchargers also lower powerplant noise as the result of taking energy from the exhaust gases to drive the compressor turbine. In addition to the speed and over-the-weather altitude safety advantages offered by turbocharging, the accompanying reduction in noise permits less fatiguing travel. By producing full-rated power at sea level without the usual 5 to 7 percent induction system losses, the turbocharger improves takeoff and climb performance at all altitudes, making the airplane less sensitive to density altitude conditions. As shown by Figure 5-7, a 210-hp turbocharged engine offers more power above 5500 ft than the sea-level-rated 250-hp engine. Unfortunately, turbocharged engines cost more initially and to maintain, require more pilot attention than unblown engines, and are usually heavier when installed. So while they are not for everyone, turbochargers do provide increased operating safety and performance over the entire flight profile from takeoff to landing.

PROPELLERS AND FANS   A constant-speed propeller improves operating performance by providing optimum thrust during takeoff, climb, and cruise. This is realized by automatically varying blade angle with airspeed to maintain a selected power setting. An engine-driven governor is adjusted by propeller pitch control to monitor rpm, while engine power is regulated by the throttle and mixture controls as indicated by manifold pressure. Although somewhat more sophisticated, heavier, and expensive than fixed-pitch, a constant-speed propeller offers reduced take-off run and time plus superior climb rate—and these are important safety items. Figure 5-8 presents the comparative relationship between fixed-pitch and constant-speed propellers, with additional discussion provided by Ref. 5.5.

Reducing diameter will lower propeller noise levels by decreasing tip speed, but something has to give to absorb the power available. This is usually realized by making blades wider or by adding another blade, both of which add to propeller weight and cost, frequently with some loss in efficiency as well. One emerging solution to both the sound level and propulsive efficiency problems is the use of a highly developed fan instead of a conventional two or three-bladed propeller.

In this country, Hamilton Standard has been the leader in research studies and evaluation of the prop fan concept. Known as the *Q-Fan* ("Q" for quiet), this propulsion system involves careful marriage of a special propeller and

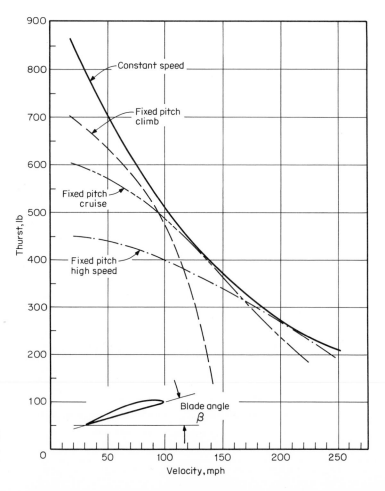

FIGURE 5-8
Propeller thrust vs. velocity. Constant-speed and fixed-pitch types.

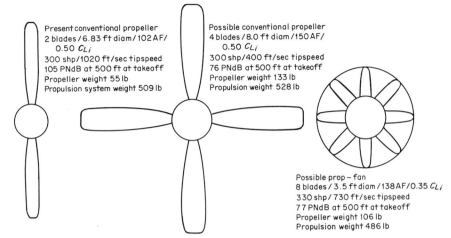

Present conventional propeller
2 blades / 6.83 ft diam / 102 AF /
0.50 C_{Li}
300 shp / 1020 ft/sec tipspeed
105 PNdB at 500 ft at takeoff
Propeller weight 55 lb
Propulsion system weight 509 lb

Possible conventional propeller
4 blades / 8.0 ft diam / 150 AF /
0.50 C_{Li}
300 shp / 400 ft/sec tipspeed
76 PNdB at 500 ft at takeoff
Propeller weight 133 lb
Propulsion weight 528 lb

Possible prop – fan
8 blades / 3.5 ft diam / 138 AF / 0.35 C_{Li}
330 shp / 730 ft/sec tipspeed
77 PNdB at 500 ft at takeoff
Propeller weight 106 lb
Propulsion weight 486 lb

FIGURE 5-9
Cessna 210J sound-level reduction possible at the same thrust by reducing propeller rpm or using small-diameter propeller fan.
Hamilton Standard via NASA CR 114289

cowling with a standard aircraft engine. Figure 5-9 indicates the diameter and noise reduction possible, as well as the comparative powerplant weights for a proposed Cessna 210J installation.

While Figure 6-6 presents a single-engine pusher adaptation of the Q-Fan powerplant, Figure 5-10 is a study of a similar quiet propulsion pod designed for light twin-engine aircraft. The small diameter and compactness of this design are quite evident, while the reduced sound levels inside and outside the cabin are most welcome. As an additional safety feature, it would be most difficult to walk into or otherwise be injured by the shrouded propeller.

A similar installation could be developed for a tractor propulsor design locating the propeller ahead of the engine. This approach has been selected for our preliminary design of Figure 6-18 in order to provide superior engine cooling during ground operation. Note the mufflers housed inside the engine cowling of Figure 5-10 to further reduce powerplant noise.

As will be noted in the propulsor fan data of Figure 5-9, the Q-Fan requires

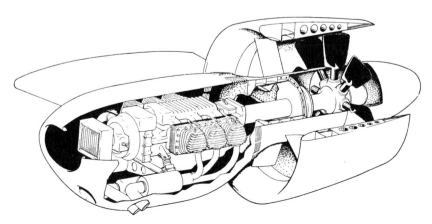

FIGURE 5-10
Q-Fan powerplant pod pusher installation.
Hamilton Standard via NASA CR 114665

about 10 percent more power than a conventional propeller and engine installation. As a result of this in combination with increasing fuel and engine costs, plus the lack of adequate experimental and advanced development capital so common throughout the entire general aviation industry, the Q-Fan powerplant has not been retrofitted to previously certified aircraft models. However, this propulsion system has a number of advantages and should be considered for new aircraft. Dowty is flying a somewhat different version of this concept in England and has been able to considerably reduce noise levels compared to a standard Britten-Norman BN-2A powerplant installation.

Our design of Figure 6-18 will use this type of powerplant to reduce propeller noise, permit superior forward visibility from the cabin, provide protection from the propeller, and reduce thrust offset distance from the airplane c.g. Readers interested in further study of the intensity and effects of aircraft propeller noise will find test data and related reports in Ref. 5.6.

While not new to the industry, Dynafocal engine mounts considerably reduce the transmission of engine vibration into the airframe and cabin area. Dynafocals, as they are known in the design office, were developed by Lord Kinematics and are a sophisticated combination of synthetic rubber and metal formed into discs which are bolted between the engine case and the aircraft engine mount. Because they are inclined so their mounting axes center at or just ahead of the engine-plus-propeller c.g., they "focus" upon the center of engine dynamic motion; hence the trade name "Dynafocal." In addition to preventing pilot fatigue, Dynafocal mounts also reduce engine and airframe structural fatigue, and so are a doubly welcome safety item.

ENGINE SAFETY INSTRUMENTS

In closing this chapter, we should note the safety advantages of cylinder head, exhaust temperature, carburetor icing, and voltage monitors. While most of us have flown for many years without these instruments, they are useful from both the operating efficiency and safety standpoints, having now been developed to a point where their advantages remove them from the "gadget" category.

While cylinder head temperature may be monitored on only one cylinder that the manufacturer found hottest during certification flight tests, it is advisable to have selective temperature indication for all cylinders. With this protection, it is possible to check temperatures to be sure there is no mixture distribution problem or air blockage to any cylinder. However, note that cylinder head temperature should not be used to determine mixture settings because of the lag in temperature response; only the exhaust gas temperature (EGT) indicator permits proper fine tuning for mixture control.

EGT indication is frequently limited to one cylinder, although once the panel instrument is installed all cylinder exhausts can be read at slight extra cost. The principal advantages of being able to check exhaust temperature include: leaning to optimum mixture ratio without overheating valves, monitoring magneto operation, early indication of poor mixture distribution or plug fouling, indication of valve problems, and indication of possible cylinder head cracks. For these reasons, it is obvious that being able to monitor all cylinders for operational efficiency or mechanical deficiency is superior to checking exhaust temperature from one cylinder just for selection of optimum mixture settings. But, in view of today's fuel prices, even this feature is better than no exhaust temperature indication at all. Properly installed and used, an EGT indicating system covering all cylinders will pay for itself many times over in greater engine life and personal safety.

The third significant powerplant safety instrument is the carburetor ice detector, (CID). Developed by ARP Industries, Inc., of Huntington, New York, this device has a probe located in the carburetor throat to detect early formation of frost. A red warning light on the instrument panel advises of the smallest ice accumulation. And, surprisingly enough, this can occur when least expected, with the warning light providing ample time to apply carburetor heat before serious ice buildup develops.

At present this detector is only available for carburetor-type engines; while fuel-injected engines can develop induction ice under certain conditions, most icing troubles occur in carburetor engines—and could be prevented by installing an ARP ice detector as shown in Figure 5-11.

The CID is superior to a carburetor air temperature gauge because the temperature gauge indicates only throat temperature and not the initial formation of frost or ice, while the CID has a variable sensitivity optical probe which immediately signals the presence of frost or ice in the carburetor throat area.

Installation of the CID probe is shown by Figure 5-11(a). The CID controls and ice formation indicator light may be mounted in a standard 3⅛-in. instrument panel opening when grouped as in Figure 5-11(b), or the controls may be remotely located with only the indicator light placed on the instrument panel along with other powerplant instruments.

A voltage monitor provides early warning of electrical problems by use of a red light to show excessive generator or alternator charging plus a yellow light to warn of system discharging. While this may not sound like any earthshaking development, the advantage of being aware that the voltage regulator is out of action or that excess charging may destroy the radios or battery is very significant. And equally important, being alerted by the yellow light that the alternator or generator is going or gone should provide time to land before all radio and electrically operated systems (such as landing gear and flaps) go dead. The yellow light will also call your attention to that master switch left "on" after the magneto/starter key has been removed for the day. Thus, this monitor can prevent finding a dead battery when an important early morning flight is about to get underway.

As a parting thought, why not combine the starter and master switches with the magneto key to preclude leaving the master switch turned on when the engine is not operating. If the automobile industry can do this, why can't aircraft manufacturers?

(a)

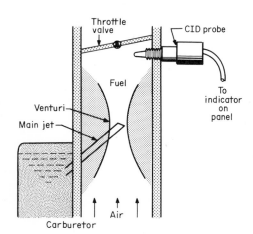

FIGURE 5-11
(a) Installation of the carburetor ice detector (CID) probe in the carburetor throat. (b) The CID control panel and probe. This panel will fit into a standard 3⅛-in. diameter instrument opening.

ARP Industries, Inc.

Since all these instruments directly contribute to flight safety, they will be in our new design—and should be important support items on every airplane used for serious flight operation.

* * *

In brief summary, the following powerplant safety recommendations are made to reduce opportunities for pilot error while improving operational safety:

1. Use an established and accepted certified aircraft engine of proven performance.

2. Select compatible engine and propeller combinations free of placarded rpm restrictions, thereby eliminating powerplant pilot error accidents caused by improper selection of rpm operating range.

3. Use pressure carburetors with automatic mixture control or Single Lever Power Control. This feature will monitor mixture settings for efficiency and required ratio, precluding pilot error through oversight or improper control position. Manual override for mixture selection should be provided for backup use.

4. Standardize powerplant controls for position, shape, color, and movement. Also standardize powerplant indicating instruments for panel location, shape, and type of readout. While standardization to this degree may be objectionable to some, pilot error powerplant accidents due to improper use of controls and fuel mismanagement will be considerably reduced.

5. Use a fuel selector valve that permits operation of all tanks as one tank for the principal setting, with individual tank selection possible for emergency or balancing purposes. The tank selector valve must have the "off" position at one end of the selection sequence. If this design arrangement is used special tank selection will not be required for any flight procedure, nor can the selector be positioned "off" during intermediate tank selection, thereby eliminating two basic sources of pilot error fuel mismanagement accidents.

6. Provide an automatically opening alternate air door in the induction system. In addition to increasing operational safety, particularly in icing conditions, this device will eliminate pilot error accidents caused by improper or late operation of powerplant controls to prevent engine failure (caused by induction system inlet icing or other blockage).

7. Develop a fuel system permitting one or more tanks to act as a single tank, thereby removing the cause of many pilot error accidents from fuel mismanagement.

8. Establish matched fuel delivery nozzle and tank filler openings to prevent filling tanks with the wrong type of fuel. Increasing numbers of powerplant accidents occur from filling reciprocating engine gas tanks with jet fuel. While this may be practical only for future aircraft models, this feature has been used for automotive (unleaded) fuels, and, had it been used in the aircraft industry, might have saved many lives over the past few years.

9. Screen fuel tank vent outlets to prevent insects from clogging the lines with mud nests, etc. This detail also provides slightly increased lightning protection.

10. Provide ample fuel for IFR and practical long distance legs—preferably at least 5 hr total, including reserve fuel. This feature should eliminate the source of many pilot error fuel mismanagement accidents caused by running tanks dry to complete the last few miles of a flight.

11. Calibrate fuel tanks to read "empty" when at least 2 gal of usable fuel remain—for the same reasons as given in item 10 above.

12. Design engine cowlings to prevent trapping any fuel from overpriming or flooding; backfire starts can cause a powerplant fire if fuel is trapped anywhere within the cowling.

13. Run the oil breather line within the passage of warm engine overboard air. This will prevent breather blockage caused by vent line moisture freezing when exposed to cold outside air.

14. Use exhaust system mufflers to reduce exterior and interior noise levels. Reducing noise reduces pilot fatigue along with the opportunity for various different but fatigue-related pilot error accidents.

15. Develop reduced-diameter propeller fans to lower noise levels both inside and outside the airplane.

16. For higher-performance aircraft, use constant-speed propellers to reduce takeoff run and increase rate of climb, two safety items that will lower the number of density altitude pilot error accidents while improving flight performance.

17. Use Dynafocal engine mounts to reduce engine vibration transmitted into the airframe and the cabin—a detail that will increase airframe life while reducing pilot fatigue on long flights.

18. Include cylinder head temperature, exhaust temperature, carburetor icing, and line voltage monitors as required components of the powerplant installation. These instruments will increase engine life and at the same time reduce the number of pilot error accidents caused by improper powerplant operation.

19. Include the master switch with the starter and mag key to turn off electrical circuits (and save the battery) when the engine is shut down, similar to automotive practice.

Having completed this review of powerplant safety features, we shall next consider airframe and equipment design details capable of further reducing opportunities for pilot error as well as improving operational safety.

REFERENCES

5.1 Harold D. Hoekstra and Shung-Chai Huang, *Safety in General Aviation*. Flight Safety Foundation, Inc., Arlington, Va., 1971.

5.2 FAA Flight Standards Service, AC No. 20-7P: *General Aviation Inspection Aids,* Supp. 5, January 1978, p. 59.

5.3 Peter E. Viemeister, *The Lightning Book.* Doubleday & Company, Inc., Garden City, N.Y., 1961.

5.4 FAR Part 23, paragraph 23.1121 (a) as interpreted by the FAA based upon prior practice.

5.5 David B. Thurston, *Design for Flying.* McGraw-Hill Book Company, New York, 1978, chap. 11.

5.6 John S. Mixson, C. Kearney Barton, and Rimas Vaicaitis, Investigation of Interior Noise in a Twin-Engine Light Aircraft, *Journal of Aircraft,* vol. 15, April 1978, pp. 227–233.

DETAIL DESIGN FEATURES

While many books could be written about aircraft detail design, our interest will principally focus upon features reducing pilot error.

As we are all aware, many automobile accidents occur daily, varying from minor fender dents to fatal wrecks. Because of the vast number of cars on the road, the actual serious accident total represents only a small percentage of all active vehicles. This results in a very low automotive fatality rate when compared to the annual mileage driven in the United States.

Because the automobile is limited to two-dimensional movement, it is restricted to ground operation. An airplane, of course, is capable of moving in three dimensions, the added dimension being vertical displacement above the earth. This vertical freedom is accompanied by the possibility of descending as a free-falling body if something goes wrong during flight. In addition, modern small aircraft can move through the air at speeds considerably greater than cars ever travel over public roads. And with stalling speeds normally exceeding the national driving limit of 55 mph, every airplane becomes a reservoir of considerable kinetic energy when in motion.

In view of this, it is quite understandable that aircraft can be seriously damaged when forced into an off-field landing, or totally destroyed by flying into a mountainside shrouded by clouds or dark of night.

While powerplant problems are responsible for a great percentage of all accidents as discussed in the previous chapter, an almost equal number result from collisions of various types. Figure 1-3 indicates that recent accidents in the categories of "powerplant and related systems failure" and "collision with ground, etc." compare within 1 or 2 percent; adding in the nose-down and nose-over accident percentages brings the total of collision accidents consistently ahead of powerplant failures. A real design challenge lies in reducing the number of collision accidents, particularly during ground operation where most occur and speeds are relatively low.

Note from Figure 1-3 that recent percentages of collision accidents are considerably below the total for 1948. This improvement resulted to a large degree from a reduction in ground accidents due to the increased visibility and superior ground handling offered by tricycle landing gear introduced during the 1950s. The nosewheel has made nose-over/nose-down and rollover accidents much more difficult to accomplish, although a few pilots still attain this goal each year. As an additional benefit, the nosewheel provides protection for the propeller and engine, and so reduces the number of powerplant damage accidents. The fact that much of this gain has been achieved by one new design feature—tricycle landing gear—proves quite conclusively that accident levels can be lowered through modification and improvement.

Let us first review what can be done to further reduce the number of collisions as well as increase the occupant's chances for survival when a crash does occur.

GROUND OPERATION

IMPROVING VISIBILITY

Starting at ground level, the first thing we need is better visibility forward and down over the nose. Taxiing aircraft still run into parked airplanes, fences, ditches, dirt banks, potholes, snowbanks, automobiles (presumably parked), hangars and other buildings, telephone poles, trees, and just about anything else likely to be found around an airport, indicating that we still have a way to go toward improving the ground operation safety record. Part of this gain could be realized by offering better visibility forward, sideways, and down, similar to the nose and cabin treatment of the Explorer landplane shown in Figure 6-1. Designed by the author in 1970, this four-place airplane provides helicopter visibility during all phases of ground and flight operation, and to such a degree that it is now being used for border patrol work in Africa.

When taxiing, the Explorer offers a view of the runway about 10 ft ahead of the nose (Figure 6-2) as well as an unobstructed horizontal field of view covering more than 280°. In flight, the view could be improved only by riding a

FIGURE 6-1
The Patchen Explorer is a flying observation platform.
Marvin Patchen Photo

FIGURE 6-2
Runways come indoors when viewed from the Explorer cabin.

Marvin Patchen Photo

magic carpet. In addition to permitting routine spotting of ground obstructions and runway potholes before reaching them, such visibility greatly increases flight safety during cruise and when in the pattern. If more of our general aviation and transport aircraft offered this level of operational visibility, the collision rate would soon post another decrease through reduction of both ground and in-flight accidents.

But these gains carry some penalty. For example, many pilots do not trust the structural attachment of a high-thrust-line engine selected here to improve forward visibility, while the unconventional configuration may have some flight characteristics all its own. And any unusual airplane causes a bit more trouble when demonstrating compliance with FAA certification requirements.

OFFSET THRUST LINE With proper design, the high-thrust-line-mounted engine can be expected to remain in position through any impact the pilot can survive. The Colonial Skimmer, Lake, and Teal amphibians, as well as the Explorer, have engine mounts designed to carry about a 15-g ultimate forward load—a value close to the maximum a belted body can accept. With the use of shoulder harness required, any future raised engine mount structures should probably be designed for 20-g maximum forward loads. Given some thought, it becomes apparent that during a crash impact exceeding 15 g the engine and propeller will continue in motion by leaving the airplane like a Roman catapult missile. This can happen—and I've seen crash results to prove it. While the occupants were

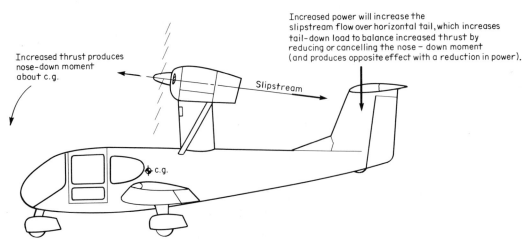

Increased power will increase the
slipstream flow over horizontal tail, which increases
tail-down load to balance increased thrust by
reducing or cancelling the nose – down moment
(and produces opposite effect with a reduction in power).

Increased thrust produces
nose-down moment
about c.g.

Slipstream

c.g.

FIGURE 6-3
The effect of a T tail and propeller slipstream in balancing nose-down moment from high-thrust-line power changes.

killed, their deaths were due to impact, not from the engine or propeller which first came to earth some 200 yd beyond and clear of the wreckage.

T TAIL   In addition, a T tail properly positioned behind a high-thrust-line propeller has the capability of balancing thrust forces so very little or no trim change results from a power change. This balance is due to increased airflow passing over the horizontal tail; as power is increased, the accompanying nose-down thrust moment is balanced by increased tail down load from slipstream flow as shown by Figure 6-3, and vice versa as power is reduced.

This same slipstream force can also be used to develop an airplane capable of climbing out of a missed or balked approach at gross weight and full power even with the stick full back. All that is required to make such a design stall/spinproof would be restricting up elevator plus interconnecting the rudder pedals with aileron control so that these surfaces cannot be crossed or the rudder displaced to start a spin.

Similar stall characteristics were realized with the original Teal amphibian designed by the author in 1968. This was the first airplane with a T tail certified by the FAA under FAR Part 23 (see Figure 6-4). Thus we find that a high engine can be expected to remain on the airplane if properly stressed, and that such unusual configurations as the Teal and Explorer aircraft can be made to fly well. In addition to being FAA certified the Teal was then custom-produced in limited volume, demonstrating that superior visibility with accompanying reductions in all sorts of pilot error collision accidents can be made available any time that manufacturers choose to produce an airplane of rather unconventional configuration.

PUSHER DESIGNS   If pusher designs are developed with aft-located engines, it is possible to have
AND FANS   the thrust line pass through the c.g. This arrangement also provides minimum

FIGURE 6-4
The Teal amphibian in flight showing the relationship of the thrust line and horizontal tail to balance power changes in flight.

Schweizer Aircraft Photo

trim change with power settings, while offering the advantages of improved forward visibility plus a quieter cabin due to the remotely located engine and propeller. However, this represents another unconventional configuration as shown in Figure 6-5, with the disadvantage and cost of propeller drive shafts and bearings necessary to position the engine sufficiently forward for airplane balance. In addition, centerline-mounted pusher propellers tend to require long and heavy landing gear struts, while the blades are quite vulnerable to tools left in the engine compartment, objects thrown or drawn from the cabin, and runway debris tossed back from the wheels.

Many of these objections to pusher-type aircraft could be overcome by using a fan-type powerplant installation originally developed by Hamilton Standard. The basic concept is that of a many-bladed propeller driven by a standard reciprocating engine. The small-diameter fan is quieter than a large propeller as the result of lower tip speed, while the shrouded blades are more efficient and less likely to strike bystanders, birds, or objects released from the cabin and cowling. Of course, there are some disadvantages associated with this improvement, mainly cost and weight as noted in Chapter 5, but continuing development work is reducing these penalties. A pusher configuration proposal using the Q-Fan installation is shown in Figure 6-6 as based upon Ref. 6.1. The improved visibility forward is obvious—and of course is our initial reason for studying these different aircraft configurations.

FIGURE 6-5
A modern homebuilt airplane, the Taylor Mini Imp, uses a pusher configuration to pass the thrust line through the center of gravity.

Taylor Photo

SHOULDER HARNESS
AND SEAT BELT

Since collisions will occur despite all efforts to the contrary, the proper place-ment and use of shoulder harness restraint systems plus well-designed crash-re-sistant structures will combine to reduce the annual fatality rate. A typical water surface collision accident is presented by Figure 6-7 showing a con-densed Teal amphibian.

Thanks to the use of shoulder harness in combination with a hull forebody designed to withstand a 20 g impact, the pilot was extracted from this wreck with no injury more serious than two broken ankles caused by his feet going through the floor boards and striking the hull bottom.

This accident occurred in Norway when a VFR pilot was flying toward the head of a fjord at dusk. He suddenly encountered surface fog which had not been apparent because of rapidly fading daylight. Rather than risk turning back with limited (VFR) instrument training, and not knowing when he would run into the mountainside at the head of the fjord, he chose to push the stick for-ward to find the water surface, which he certainly succeeded in doing, thereby proving conclusively that at over 50 mph the water surface is as hard as any ground. This airplane has since been rebuilt and is flying again with the same pilot in command—but he now has an IFR rating.

Shoulder harness has been used in all military aircraft since the beginning of World War II, and has been standard in agricultural aircraft for many years. Based upon the studies of Ref. 6.2, the human body can withstand fore and aft deceleration of 45 g for 0.10 sec applied either to the chest or back, while lat-eral deceleration (left to right or vice versa) up to 11.5 g can also be applied for 0.10 sec without permanent injury. Vertical acceleration along the spine re-sulting from the vertical component of a crash landing can be as high as 10 g

FIGURE 6-6
Aircraft design using a Q-Fan pusher powerplant for im-proved visibility and reduced noise level. (a) Possible Q-Fan powerplant pusher configura-tion. (b) Q-Fan single-engine pusher installation.

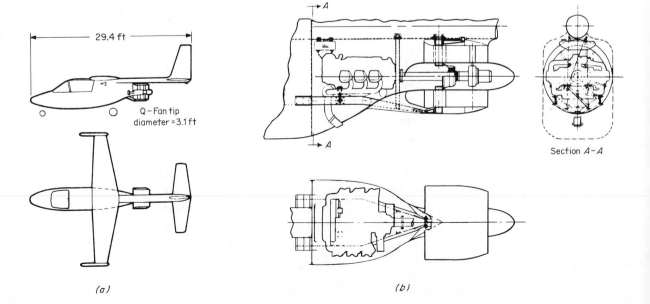

FIGURE 6-7
A Teal amphibian force-landed in the water to avoid the head-wall of a Norwegian fjord. Because of the crash-resistant design and shoulder harness, the only injury to the pilot was a pair of broken ankles.
Photo by Inspector Smith—Overland Norwegian Directorate of Civil Aviation

for 0.10 sec—a level we can realize with crash-resistant structures and improved seat cushioning materials.

As of July 18, 1978, the FAA required the front seats of all newly manufactured general aviation aircraft to be equipped with both shoulder harness and the usual seat belt, while rear seat passengers must have at least a seat belt. In Great Britain the CAA required a safety harness (or diagonal shoulder strap) plus seat belt for all passengers in registered aircraft after January 1, 1978. Similar shoulder harness plus seat belt equipment has been mandatory in Sweden since 1970 and in Australia since 1973.

Unfortunately, a poorly designed or positioned shoulder harness may be worse than none at all, regardless of all the government legislation. It is essential that the shoulder harness be anchored to withstand an impact force of at least 40 g for a 100-lb upper torso weight, which becomes a 4000-lb attachment limit design load for each passenger being restrained. This load is not easy to carry into the structure and still have the harness properly led across the torso. For example, in a four-passenger or larger cabin, where can the upper end of the harness be attached without getting in the way of rear seat passengers? If led to the ceiling, it is at too high an angle to prevent body stretch and head damage.

Further, many diagonal single-strap shoulder harnesses are led down and inboard across the torso to such a low point that the torso can roll out of the harness upon initial impact. To prevent this, a dual shoulder harness is no doubt superior to a single diagonal strap (and would be for automobiles as well); in fact, to make sure the pilot's torso can neither go up nor forward, most agricultural (Ag) planes have dual shoulder harnesses that are tightly strapped down without including any inertia reel. Of course, a pilot of small stature may have difficulty in reaching all controls with this restraint system, so improved types are coming into use with a Y-shaped harness going over the pilot's shoulders and leading aft to a single 4000-lb-strength inertia locking reel. The reel is then mounted onto a structu e capable of distributing a load of this size.

That is the best system for all general aviation aircraft. Dual straps leading from the midpoint of the seat belt should be brought over the shoulders and joined to become a single strap. This strap then leads aft and outboard to a 4000-lb-capacity inertia locking reel anchored onto a sufficiently rugged side longeron of the fuselage. The strap should go aft from the shoulders approximately horizontal as shown in Figure 6-8.

If possible, structural attachment of seat belts and harnesses should be designed to slowly yield (deflect) under load starting at around 15 to 20 g (1500- to 2000-lb total harness load and 750 to 1000 lb at each end of a seat belt). This structural deflection reduces sharp peak loads and tends to even out the distribution of load on the human torso. The S.A.G.E. Corporation of Burlington, Vermont, has developed a force-limiting seat belt anchor that deflects under a 20-g load, but the device, known as a Dynastat, is not presently available for small aircraft use. Hopefully, it will be soon.

These belt and harness features should not be limited to the forward occupants, but should be available for every seat in the airplane. Pilots would be more inclined to use a shoulder harness if it contained a positioning clip similar to automobile harnesses to take the retraction reel load from a pilot's shoulders. The constant pressure of presently available personal aircraft harnesses becomes such a nuisance on long trips that the harness is often ignored. But when the harness is suddenly needed, there is usually little opportunity to strap it on; so increased safety would be realized by making comfortable shoulder harnesses for normal, routine flight because they would be used more often.

Downward forces up to the 10 g for 0.10 sec that can be withstood without injury (Ref. 6.2) can be absorbed by a new cushioning material originally developed by Edmont-Wilson for wheelchair and bed patients. Marketed by Commonwealth Cushions, Ltd., of Yorktown, Virginia, under the trade name "Temper Foam," this material comes in various densities for proper load distribution. Although these cushions feel stiff when first sat upon, the synthetic foam yields and flows to conform to body contours, providing a most comfort-

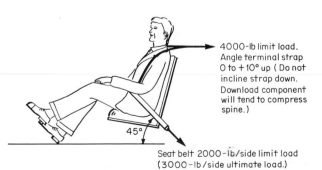

4000-lb limit load.
Angle terminal strap
0 to +10° up (Do not
incline strap down.
Download component
will tend to compress
spine.)

45°

Seat belt 2000-lb/side limit load
(3000-lb/side ultimate load.)

FIGURE 6-8
Recommended lead angles and attachment loads for seat belt and shoulder harness.

able and relaxing seat. The same concept applies to sudden high-g-force impacts—the cushion compressed comparatively slowly to reduce impact by displacement without bottoming out at a sudden peak force. A mixture of different density layers seems to provide the best seat and back cushioning as recommended by Commonwealth. With 100 percent recovery after compression, Temper Foam is as durable as any polyurethane cushion.

The seat structure and attachments as well as the supporting structure under the floor should be designed to accept this 10-g down load, which is much higher than present FAA requirements of 3 g (Ref. 6.3). Typical belt and harness installations are shown in detail in one FAA publication (Ref. 6.4), although the maximum 30° aft-and-up slope of the harness reel strap shown in that data is considered too high (see Figure 6-8 for preferred angles and loads).

CRASH-RESISTANT
DESIGN

Crash-resistant structures can do much to save lives, as shown by Figure 6-7. Designing a fuselage to withstand 20-g head-on impact is not an impossible, overly heavy, or expensive task. In fact, the Schweizer 1-35 high-performance sailplane is designed to absorb a 20-g impact along the cockpit area, with further yielding at larger values; I personally know of one life saved by this feature.

Heavier aircraft, of course, require more structural weight to provide this cabin protection than do light aircraft. Contrary to our forward visibility preference for engines located somewhere other than in the nose of an airplane, having the engine forward to impact first takes a large percentage of aircraft weight out of the crash condition. An aft-mounted engine will increase impact forces and tend to come forward; however, a raised-thrust-line engine will separate catapult style at severe impact as noted and so reduce the total impact force. In similar fashion, wings can be designed to collapse forward at 15 to 20 g by having each panel rear attachment shear off; the resulting immediate weight reduction decreases the impact acceleration force accordingly.

As shown by the curves of Figure 6-9, the g force of any crash decreases with speed at the moment of impact and/or with increased stopping distance. If contact can be made at a speed as low as 40 mph, the force from stopping within 3 ft would be less than 19 g compared to 41 g for stopping in the same distance from 60 mph. This is a plus for having low landing speed in any airplane design, but be careful of slowing up prior to emergency contact to such a degree that the final phase is a spin from low altitude. Thus, we find another good reason for the increased safety of a spinproof airplane.

Since aircraft do not normally crash either vertically or horizontally but at some angle of descent, let us consider the impact of a forced landing at a 10°-angle approach path and 60-mph glide speed. The vertical rate of descent will be 10.4 mph (60 mph × sine 10° = 60 × 0.1736 = 10.4-mph vertical descent rate). If vertical acceleration can be stopped within 1 ft, the maximum vertical impact will be 4 g (see Figure 6-9). This means that an aircraft fuselage

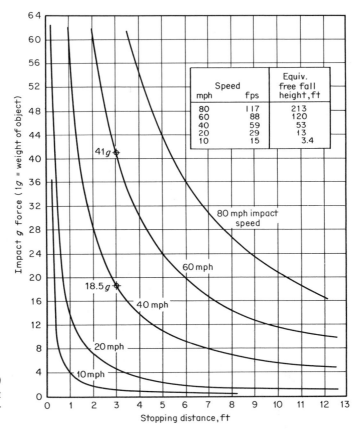

	Speed		Equiv. free fall height, ft
mph		fps	
80		117	213
60		88	120
40		59	53
20		29	13
10		15	3.4

FIGURE 6-9
Stopping distance and impact speed vs. impact g force developed (uniform deceleration).

bottom structure capable of collapsing for 12 in. without crushing legs, rupturing fire-producing fuel lines, or pushing lethal fragments into the cabin would limit the vertical component of a 60-mph 10° glidepath to a 4-g impact. When you consider that the average instrument approach path is around 3°, a 10° descent glide seems reasonably steep for controlled emergency descent.

Of course, the forward component of velocity would still produce a fore and aft impact load—but note that if this landing was made in comparatively clear terrain or brush that yielded upon contact, the horizontal impact force would be less than 11 g provided the horizontal stopping distance after contact exceeded 12 ft.

Increased crash protection is one major reason for hull amphibian safety; the strong bottom required to support water loads not only resists fairly rugged terrain, but also has some distance to go from the keel to the cabin floor to absorb impact energy. Landplanes could be designed with similar protective features for less than a 10 percent increase in fuselage weight, which would result in about a 1.5 percent addition to the airplane empty weight. This premium seems justified for the increased safety provided during emergency landings or unexpected surface maneuvers such as ground and water loops.

GROUND AND WATER LOOPS

Despite almost universal acceptance of tricycle landing gear, *ground loops,* which are really high-speed uncontrolled turns, continue with surprising frequency. Some of these result from poor cabin visibility requiring last-minute evasive action, such as swerving to avoid a collision, which ends up in a ground loop instead. They also occur as secondary accidents following aborted takeoffs caused by powerplant failure or density altitude conditions, overshooting or undershooting landings, and landing gear failures during takeoffs or hard landings.

A water loop accident can be more serious than the usual ground loop since the seaplane may suddenly dig into the surface and then turn with such extreme rotational acceleration that occupants can be thrown out at high speed or crushed against the cabin interior. Because the water surface can act like solid ground and damage structure accordingly, a water loop frequently causes more extensive damage than a ground loop.

The annual tally for ground and water loop accidents remains fairly steady at 13 to 15 percent of all first types of accident as noted in Figure 1-3. This amount of damage is worthy of serious analysis, with some design consideration directed toward reducing high-speed turning accidents; but let us first see how and why they happen.

During land operation, most ground loops occur with so-called conventional (tailwheel-type) landing gear. A fair percentage of such accidents are caused by pilots who have been trained in tricycle gear aircraft and have little or no conventional gear experience, which seems to indicate that some inherent problem exists in the conventional landing gear configuration—and indeed there does.

As shown by Figure 6-10, any tailwheel gear is basically unstable. When all parts of the airplane are headed along the runway in the same direction and into the wind as well, the tailwheel will stay in line where it belongs. But if a crosswind gust displaces the tail or a landing is made with some skid or crab angle present at contact, the weight of the airplane may (and usually will) cause the airplane to ground loop as shown in Figure 6-10(*b*).

To make matters worse, the further forward the main wheels are placed to prevent nosing over, the larger the tailwheel loading becomes; and the heavier the tail load, the greater the tendency to ground loop, as will also be noted from this illustration.

Fortunately, most tailwheel aircraft are equipped with a steerable wheel that remains engaged until the rudder has been turned about 30° off the airplane centerline. This provides lateral resistance to sidewise motion of the tail. But unfortunately, 30° may be the amount of rudder deflection suddenly required to correct a severe skid or gust condition. So, just when tail restraint is required, the tailwheel releases into free swivel and away we go upwind as in Figure 6-10(*b*).

Various solutions to tailwheel landing gear problems have been offered over the years (Ref. 6.5), including multiple wheel systems and "self-castering"

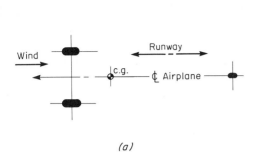

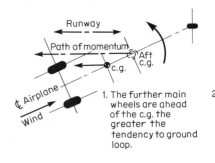

1. The further main wheels are ahead of the c.g. the greater the tendency to ground loop.

2. A free swiveling tail wheel rides with the turn and provides no corrective restraint.

(a)

(b)

FIGURE 6-10
Stability characteristics of conventional landing gear. (a) Conventional gear tracking down runway. (b) The ground loop starts. If the momentum vector gets outside the tread, a ground loop is underway. Note that the aft center of gravity requires a smaller skid angle to start the windup. Tracking correction must be made immediately.

main wheels (also known as *crosswind gear*), but the most successful to date has been tricycle landing gear. Although used on the P-38, P-39, and some other military aircraft, Fred Weick's Ercoupe really introduced this landing gear arrangement to small aircraft just before and again right after World War II.

Tricycle gear performs so well during landing runout because, as shown by Figure 6-11, the location of the main wheels behind and outboard of the c.g. makes them tend to trail directly behind the c.g. This action is called *inherent dynamic stability*. We all know it is easier to pull a two-wheeled cart uphill than push it, particularly if the tow bar is pinned to the cart. And pushing a cart with a pinned bar is similar to the action of conventional gear with a free-swiveling tailwheel.

As indicated by Figure 6-11(c), a ground loop is possible with tricycle gear if the nosewheel is turned upwind at or just after ground contact. Fortunately, the final approach procedure in crosswind conditions will normally include a bank into the wind to stop drift plus some top rudder to keep the runway centered. So even if the nosewheel and rudder steering are interconnected, runway contact will usually occur with the nosewheel aligned with the direction of motion —or at least that should be the intent. This requirement also explains why modern nosewheels have spring bungee steering arms which allow the drag load to straighten any nosewheel offset angle present at runway contact.

As you may know, neither the conventional gear tendency to ground loop nor a possible tricycle gear skid upwind are nearly so likely to occur if you are operating from a grassy field. Owing to reduced surface restraint, a grassy field

FIGURE 6-11
Stability characteristics of tricycle landing gear. (a) Tracking down runway. (b) Recovery with a free-swiveling nosewheel. Momentum acts to recover from a skid. (c) Windup possible with steerable nosewheel when holding rudder at touchdown to correct for sudden crosswind gust.

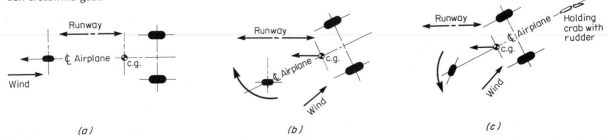

(a)   *(b)*   *(c)*

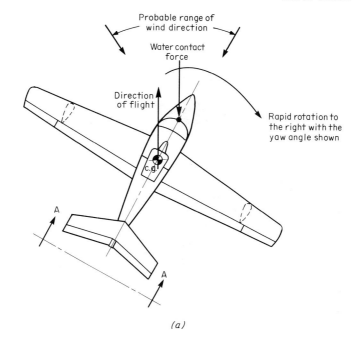

Probable range of
wind direction

Water contact
force

Direction
of flight

c.g.

Rapid rotation to
the right with the
yaw angle shown

A

A

(a)

FIGURE 6-12
The water loop: cause and effect. (a) How water contact
ahead of the center of gravity
combined with some angle of
yaw (crab or skid condition)
provides the basis for a water
loop. (b) View A-A, showing
how inertia and water forces
act to bury the outside wing
and float during a water loop
turn, frequently removing the
outer panel.

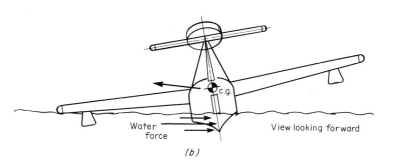

c.g.

Water
force

View looking forward

(b)

becomes a most forgiving partner during those momentary lapses of landing technique. And the nice, square, grassy fields we formerly enjoyed at various aviation country clubs and airports around the country in the late 1940s also provided every possible opportunity to land into the wind, further reducing the tendency to ground loop.

These airfields coupled with proper pilot technique partially explain why fewer ground loop accidents occurred in the days of conventional landing gear, as clearly shown by the Figure 1-3 total for 1948. The remaining differ-ence in percentages is probably due to changes in reporting procedure plus the fact that fewer seaplanes were around in 1948, so there were not as many op-portunities for water loop accidents as we have today.

Figure 6-12 diagrams the basic cause of water looping. Assuming a sudden turn is not required to miss something in the water, such as a log or suddenly appearing powerboat, water looping primarily results from excessive approach speed coupled with some yaw or crosswind drift. A high-speed approach causes any seaplane to trim nose-down, putting the initial water surface contact point well forward of the step and even possibly just aft of the bow. As clearly shown by Figure 6-12(a) the combination of inertia and water forces at the moment of yawed contact provides considerable rotational energy which is rapidly converted into angular acceleration.

By way of compounding water loop problems, if the seaplane is an amphibian with nosewheel doors, these may be sucked open at contact and act as water brakes to rapidly decrease forward velocity and increase rotational speed. With water behaving like concrete at high speed and with various inertia forces trying to upset the seaplane as shown in Figure 6-12(b), it is small wonder that the water loop can prove fatal to occupants as well as severely damage any seaplane structure. In fact, failures of the outer wing or forward hull structure are quite common in the loop, frequently sinking the seaplane. And this type of accident happens so fast it is usually difficult for participants or observers to accurately recall just what happened first.

IMPROVED LANDING GEAR DESIGN

With this "surface looping" background, we can now consider landing gear improvements that will reduce the extent and frequency of structural failures caused by unexpected ground loads, thereby lowering both pilot error and accident totals. Water loops are primarily a pilot technique problem as discussed, and a possible future design solution is noted at the end of this chapter.

To permit rough-field skid or drift landings, ski operation, and to prevent gound loop accidents, the nosewheel strut and supporting structure should be designed for greater strength limits than required by FAA specifications. The vertical, aft, and side loads in particular should be increased. A safe vertical limit seems to be obtained by designing for a braked landing with the full inertia of the airplane mass (rotating about the main wheel contact point) resisted by the nosewheel strut. This load will generally exceed current requirements by 1.5 to 2 times, and must be carried by the nosewheel support structure as well as the strut.

Side and aft load factors should be doubled to prevent nosewheel failure when an airplane strikes snow and ice ridges or is operated from rough grass fields. While these higher loads will add about 10 percent to nose gear weight and cost and slightly reduce useful load, this safety tradeoff more than pays for itself the first time the gear does not fail when it might have; this is especially true for bush-country operation. By incorporating stronger nosewheel structures, we can prevent nose gear failures that are usually charged to pilot error. In addition, the nosewheel must be steerable for ski operation because brakes and skis don't mix.

Another source of nosewheel failure is *wheelbarrowing*, a term applied to high-speed landings during which initial ground contact is made on the nosewheel, causing an airplane to look like a wheelbarrow going down the runway. Frequently the landing continues as the propeller blades bend back after the nosewheel strut collapses—with the entire airplane finally coming to rest on the cowling bottom or upside down facing backwards (nose-over accident).

The solution to this problem lies in designing the airplane with the thrust line essentially parallel to the ground line. The thrust line is mentioned rather than the wing angle of incidence (the angle between the wing chord line and the fuselage reference line), which is the real culprit, because the thrust line is usually about level in cruising flight. When speed is reduced below cruising velocity, the wing requires a greater angle of attack (the angle between the wing chord line and the relative airflow) to realize the increase in lift coefficient necessary to support the airplane at a slower speed. This, of course, raises the thrust line—and if the design ground line is parallel to the thrust line (or airplane cruising trim reference line) it follows that the nosewheel will be clear of the runway (higher than the main wheels) at any speed below cruising speed as shown by Figure 6-13.

This design feature will prevent nosewheel collapse due to wheelbarrowing, because an airplane cannot land on the nosewheel unless landing contact is faster than cruising speed. Since this is highly unlikely, we have removed another cause of pilot error landing accidents. A landing gear parallel to the thrust line should also decrease takeoff run and time because of reduced airplane frontal area, which means lower drag during takeoff acceleration.

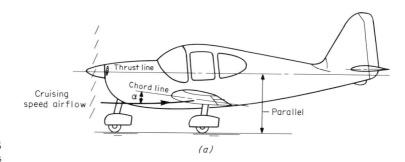

FIGURE 6-13
Tricycle gear with wheels parallel to thrust line prevents landing on nosewheel strut prior to main gear contact. (a) Thrust line, airflow, and tire bottoms parallel for cruising flight. Thrust line and tire bottoms parallel at rest on runway (α is angle of attack required to develop lift at flight speed). (b) Increase in angle of attack at landing speed provides nosewheel clearance from runway, preventing "wheelbarrowing" type nosewheel structural failure. (Note angle between thrust line and ground.)

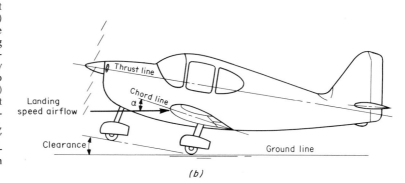

Main gear struts collapse from taxiing into holes or similar obstructions such as ditches and ridges, drift or skid landings, bounce landings that become hard landings, and ground-looping after touchdown. These conditions can impose high side or vertical loads on the main strut and its supporting structure—and when the gear collapses, we have another pilot error landing accident on the records.

For about a 10 percent main gear weight and cost penalty plus some reduction in useful load similar to the nosewheel installation, it is possible to increase main gear side and vertical load capability by about 50 percent. The major items requiring additional strength will be the wheel axle, side load strut, pivot axis of retractable gear, and the local supporting structure.

Naturally, regardless of the strength added, situations resulting in gear collapse will still occur, but the frequency of routine ground accidents would be immediately reduced. The effect upon insurance rates would also be favorable, so the goal again seems doubly worthwhile.

The use of tricycle landing gear will add about a 30 percent increase in landing gear weight and cost compared to tailwheel gear, while increased nose and main gear design load factors will add another 10 percent as noted. The net result is a 40 percent increase in fixed tricycle landing gear weight and cost compared to conventional gear. This means that a standard tailwheel-type landing gear of 100 lb, costing $1000 to build and install, would translate into a 50 percent stronger tricycle gear weighing 140 lb and costing $1400. The 40-lb difference is just over 6½ gal of fuel or other equivalent useful load, and seems a small penalty for the added safety and utility provided. In addition, because of the reduction in landing gear ground accidents, lower insurance premiums should compensate for the added first cost within 2 or 3 years.

In summary, by designing our aircraft with tricycle gear of greater strength than required by current FAA specifications, we can reduce the frequency of pilot error–induced landing gear failures resulting from collisions and improper operation.

The subject of landing gear design detail would not be complete without a word concerning retractable landing gear. Since owners of retractables seem to be divided into two classes, those who have landed gear up and those who eventually will, the possibility of improving fixed-gear aircraft performance is worthy of development study.

Figure 6-14 data for fixed gear with unfaired and fully faired wheels and struts indicates proper fairing may provide more than half the improvement possible with fully retractable gear. In addition to the saving in weight and cost possible with fixed gear, it is obviously impossible to land gear up if the gear does not retract. This worthwhile approach may not be practical for aircraft capable of cruising over 170 mph, but it has merit for thousands of airplanes now offering around 150-mph cruise performance with retractable gear. For proof that fixed gear need not preclude a respectable cruising speed, consider Piper's Archer II cruising at 144 mph with 180 hp, the 235-hp Dakota at 158 mph, and

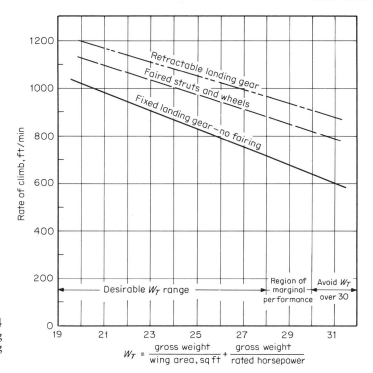

FIGURE 6-14
Landplane (tricycle landing gear) climb rate vs. landing gear fairing and W_T.

$$W_T = \frac{\text{gross weight}}{\text{wing area, sq ft}} + \frac{\text{gross weight}}{\text{rated horsepower}}$$

the Cherokee Six cruising 153 mph with 260 hp—all with fixed, faired tricycle landing gear as typically shown in Figure 6-15.

If landing gear must be retractable for maximum performance, the automatic extension system used by Piper and Bellanca to lower the gear at low approach speeds should certainly be used. This excellent feature has saved many pilots

FIGURE 6-15
Piper Warrior II (bottom) and Archer II (top) showing fixed landing gear streamlining for improved performance.
Piper Aircraft Photo

much embarrassment as well as being favorably reflected in lower insurance rates. But when incorporating the automatic gear extension into any design, there should be no method of cutting out or bypassing this system. Some aircraft include a manual override to permit gear retraction for maximum glide ratio in case of powerplant failure; as a result, pilots go up to practice low-speed flight and cut out extension operation to prevent the gear from going up and down with speed changes. It would seem logical, however, to practice low-speed flight and stalls with the gear extended since the gear is supposed to come down when making an approach at correct speed. Frequently, the net result of flying at altitude with automatic extension inoperative is a wheels-up landing. This happened twice within 2 months to the same airplane at a nearby airport (one of the months being required to repair the airplane after the first accident); each time the deed was done by a different pilot who had been practicing low-speed flight at altitude with the gear extension manually overridden. Conclusion: the automatic landing gear extension feature is most desirable for retractable-gear aircraft, but for safety a bypass should not be included and does not appear necessary.

Following this line of thought, the use of a "squat switch" is also a highly recommended safety feature. Squat switches are normally closed switches located on all three retractable landing gear struts. When weight is on the oleo struts or some part of each gear leg displaced by airplane weight, the switches are forced to the open position, breaking the retraction command circuit and so preventing retraction of the landing gear. As individual struts clear the ground and fully extend during takeoff, all three switches become closed, permitting operation of the gear retraction system. With this simple circuit, our final landing gear safety feature, it is impossible for a pilot to commit the error of selecting "gear up" before becoming fully airborne. Of course, gusty air can drop an airplane back onto the runway after becoming fully clear, so some judgment and consideration of takeoff conditions will still be necessary.

FLIGHT OPERATION
STANDARDIZED CONTROLS

Airplane rental is becoming increasingly common and should continue growing in the future owing to the high initial and maintenance costs associated with occasionally used aircraft. With a large number of pilots checked out and approved to fly more than one airplane, standardization of instrument and radio panel layout, switch location, and secondary controls placement becomes of major importance—particularly in emergency conditions. When stepping from one type of airplane into another, a pilot should not have to hunt around for carburetor heat, tachometer location, radio selection system (which frequently defies logic), arrangement of omni and automatic direction finder (ADF) dials, Pitot heat switch, etc. Ideally, they should be in the same location on all similar class aircraft; i.e., one standard arrangement for single-engine aircraft and another similar layout for twin-engine models.

Powerplant controls for personal aircraft should be of the same type, color, and location for all makes and models, with the possible exception of aerobatic aircraft (which have many special design requirements). For example, with standard quadrant-type controls located on the centerline, Cherokee style, the throttle knob should be black, the mixture red, propeller pitch blue, and carburetor heat yellow. This identical color coding has been successfully used in Finland for over 15 years, so it is not entirely new. All knobs should be distinctly different in shape as presently required by FAA design standards to further distinguish control separation by feel, but should be of identical shape for the same control in any airplane. The more obviously dissimilar various controls can be made, the less likely a pilot will operate the wrong one by accident, such as pulling back mixture when reduced throttle or rpm change is desired — the result, no power and a forced landing, possibly from low altitude or on final.

A color-coded checklist should also be prominent among standardization items covering takeoff, cruise, and landing on land, with landing on water included under a separate heading when applicable. The background colors could be light gray for takeoff, light blue for climb and cruising speeds — relating to being in the sky, and light orange for the landing check. When water landings are also possible as would be the case for amphibians, a fourth checklist heading would include water landing items carried on a light green background — to associate this check with water. For floatplanes, light green for water would replace the light orange landing checklist. Light colors are suggested because black lettering of the various checklist items would show clearly and permit easy reading. Of course, no checklist is of value unless referred to as a regular part of each segment of flight procedure.

This or a similar cockpit standardization program should be coordinated either by the General Aviation Manufacturers Association (GAMA), the Aircraft Owners and Pilots Association (AOPA), or a new organization formed for the purpose before the FAA or some other government bureau legislates such requirements. Considering the number of accidents attributable to pilot mismanagement of secondary controls and misinterpretation of instruments, the safety contribution of cockpit standardization would be very real and is long overdue. Further, this is the type of program that should be coordinated and determined by the industry without government interference.

TORQUE EFFECTS Torque effects accompanying sudden or prolonged power changes cause takeoff accidents and can also become tiring if right rudder must be held during long climbs or when practicing repeated takeoffs and landings. For example, late application of right rudder when power is suddenly applied to go around after making ground contact too far down the runway frequently ends in a left swerve off the runway due to torque effect. For our clockwise rotation engines, why isn't the thrust line offset to the right to balance these forces? Model airplane builders have been doing this for over half a century, so it is no great new development. But it is one that would help prevent many pilot error accidents. (Of course, the thrust line should be offset to the left to reduce counterclockwise rotation torque effects.)

FLIGHT INSTRUMENTS Many takeoff accidents are the result of failure to obtain flying speed, usually caused by density altitude conditions, instrument errors, overloading, or pilot technique. Solely from the design standpoint and without regard to flying ability, the simple dynamic pressure-sensing instrument discussed in Chapter 4 could prevent many of these accidents by indicating when optimum rotation speed is attained.

By sensing flow pressure this indication would automatically compensate for density altitude variations affecting wing lift, and by presenting a "go" signal within the field of vision would ensure that pilot attention is not diverted from the takeoff path. To provide the ultimate in sophistication, takeoff indication could be made adjustable for flight weight by tailoring this instrument for specific aircraft models. The price of such equipment would be in the range of existing lightplane AOA indicators, or about $500. As part of the total head-up display package, the cost would be considerably less. This device could be developed in a similar way to the speedring used in cross-country sailplanes and for about the same cost.

Incorrect airspeed indication can result from airspeed lines clogged by moisture or insects, particularly mud wasps which regularly plug fuel vent lines as well (see Chapter 5). To remove trapped moisture in airspeed lines, the cause of many erroneous speed indications, the lines should run from the airspeed head without any low spots to position their lowest point at a convenient drain in the cockpit—and, of course, each line should have its own quick drain fitting.

The solid plugging of Pitot head static and pressure pickup openings by mud wasps is quite frequent and difficult to note during a preflight walk around, because the trouble is inside the airspeed head and so out of sight. Using a flagged cover or sock over the head will work, provided the takeoff checklist notes removal of this cover prior to flight. An alternate static source is also a desirable design feature, and is required basic equipment in Canada and most foreign countries. However, if the Pitot (pressure) port has been clogged, the alternate static is of benefit only to the altitude and rate of climb instruments, since airspeed indication is based upon the difference between total and static pressures and will not work properly if either or both lines are clogged.

When discussing flight instruments, we should give consideration to different power sources for IFR gyros. A sound arrangement would include operation of directional and attitude gyros from an engine-driven vacuum or pressure pump, plus an electrically driven turn and bank indicator.

The use of an externally mounted venturi can no longer be recommended for serious IFR work, either as a primary or secondary power source, because a venturi is subject to icing in that unexpected snow or ice storm and can go out of action just when needed most. However, some success has been realized under all flight and weather conditions with a venturi vacuum source located in engine-cooling airflow at the cowling exit opening.

IMPROPER LOADING   In addition to attempting takeoff prior to attaining sufficient flight speed, pilots frequently have accidents resulting from improper loading. If we neglect overloading for the moment, since avoiding this unfortunate condition is really a matter of common sense coupled with simple addition and subtraction, the most damaging type of improper loading occurs from exceeding aft c.g. limits. This makes the airplane less stable in pitch and may result in elevator control force reversals as reviewed in Chapter 8. Moving the engine, the largest single mass item on small aircraft, away from the nose to a midfuselage high-thrust-line installation (Figures 6-1, 6-3, and 6-4) or aft for a pusher installation (Figure 6-5) results in a far aft c.g. when the airplane is empty. As a result all seats, fuel, and baggage can be located forward of the empty weight c.g.

This means that passengers, baggage, and wing leading edge fuel will bring the c.g. forward from the empty weight aft c.g. position. Since the furthest aft flight c.g. will have to be demonstrated as being safe at the minimum possible flight weight (i.e., a pilot and 4 or 5 gal of fuel plus engine oil), any additional weight added as useful load will act to move the c.g. further forward. This is a basic flight safety feature that virtually precludes loading such a configuration behind the permissible aft c.g. limit, and thereby prevents aft c.g. pilot error loading accidents. However, despite all design efforts to prevent aft c.g. accidents, we still can do nothing to help the pilot who chooses to throw on a few hundred pounds of extra weight on a 110° day, 3000 ft above sea level—and then proceeds to take off into the trees because of overloading in adverse density altitude conditions.

STOL AIRCRAFT   The takeoff and landing safety record can be improved by turning to STOL aircraft, which offer superior safety when density altitude, short-field, or both operating conditions are frequently encountered. As indicated by the W_T values of Figure 6-14, the lower the combined wing and power loading totals, the better the climb performance. The superior short-field operation of the J-3 Cub shows that great amounts of power are not necessary if the wing loading is sufficiently low, although more power for a given wing loading and gross weight means more climb. Since our small aircraft ride more comfortably and cruise faster if wing loadings are in the 14 lb/sq ft range, power loadings should be around 10 lb/hp to provide the W_T value of 24 or less necessary for minimum acceptable STOL performance of light single-engine aircraft.

The ability to rapidly climb out of small fields and ponds, or over steep hills and other obstacles is a safety feature that can frequently replace pilot judgment to reduce pilot error accidents. In addition to protecting inexperienced pilots, this capability is particularly important for anyone faced with extreme density altitude conditions.

Of course, being able to land steep and short under full control provides equally satisfying small-field and emergency landing safety. To accomplish this an airplane should have a wing loading closer to 10 than 14 lb/sq ft and be able to slip with ease under full control at low speed—and without the possibility of stalling. However, these last features are in opposition. To slip requires the ability to apply opposite aileron and rudder controls: for example, left aileron, stick back somewhat to keep speed down, and right rudder—with stick back too much, a low-altitude spin can result. But to prevent spins it is necessary to interconnect ailerons and rudder to move together, while elevator-up travel must be limited to prevent wing stall—so it will be impossible to slip.

This apparent contradiction is easily resolved by the use of wing-mounted speed brakes (or spoilers) discussed in Chapter 3. Extension of these brakes reduces or destroys lift over part of the wing to increase descent rate; and since lift is fully restored when the brakes are retracted, the airplane will return to its original glide angle. This capability provides a fine degree of glidepath and landing-contact-point control—and with some practice will be found superior to sideslipping.

This means stall/spin-free STOL approach is possible by keeping wing loading around 10 lb/sq ft, interconnecting aileron and rudder controls, restricting elevator-up travel, and using speed brakes for descent-path control. The wings could have flaps for STOL takeoff and to reduce wing area for better cruising performance, although some deterioration of slow-speed approach ability may be expected if wing loading exceeds 12 lb/sq ft.

STOL capability should materially reduce pilot error accidents due to overshooting and undershooting, while providing safer operation from small fields and airports surrounded by obstacles of any sort. As you can see, our airplane "designed for safety" is gradually taking shape by the logical process of incorporating detail features increasing operational safety.

CONTROL SURFACES   To provide maneuverability during the landing flare or in "tight" weather spots, all control surfaces should be effective right down to minimum flight speed—whether this is at wing stall or limited by elevator travel. Running out of rudder or aileron power during a low-speed control correction can be frustrating, and expensive if a collision occurs.

In order to assure maximum effectiveness, control surfaces should be sealed and located as far as practical from the airplane c.g. to provide adequate control power at all flight speeds (Ref. 6.6). Sealing between movable and fixed surfaces may be as basic as a sponge rubbing strip, but even this simple detail will reduce airflow between the high and low pressure areas on either side of a control surface and so increase lift developed by a deflected surface.

More attention should also be paid to balancing control forces between the various surfaces. That is, the aileron load for any maneuver should feel in correct proportion to the elevator and rudder forces required. Heavy ailerons are a frequent complaint from hull-type amphibian owners, but this particular unbalance is most difficult to reduce without extreme control wheel or stick travel. Aircraft designed as flying boats, whether amphibians or seaplanes, must have ailerons large enough to raise a low wing and tip float at very low speed when on the water. This characteristic is necessary to permit lateral control when coming up onto the step, usually around 25 to 30 mph, but tends to make aileron control forces high during cruising flight. However, for landplane personal aircraft, all three primary control forces should feel about the same magnitude and be sufficiently low to prevent pilot fatigue on long journeys.

In order to satisfy current FAA flight demonstration requirements, a rudder trim system is required in addition to quite powerful elevator trim. These trim controls must be designed to provide pitch and directional (yaw) control in the event the primary controls fail (presumably not from jamming of the control surfaces, since in that case the tabs could not move them either). But if an elevator or rudder cable or pushrod should separate, the ability to trim in pitch and yaw would be most welcome. Thus, these trim features would also contribute to flight safety in certain emergency conditions.

Trim tabs of built-up constructions, that is, having ribs and skin covering, should be designed to drain collected moisture. As part of preflight inspection all movable tabs should be checked for ice and mud accumulation, which could unbalance the control surfaces. Although every new design must be checked to design dive speed V_D, and every production airplane is taken to never exceed speed V_{NE} (which is $0.9V_D$), these aircraft are checked under controlled conditions and with all details exactly like the certified design data. If the surface balance changes from accumulation of dirt or ice, flutter could develop. In fact, this caution applies to all movable control surfaces; they should be frequently checked to make sure service use, repainting, and exposure have not resulted in changes in balance which might cause surface flutter.

AEROBATIC AIRCRAFT   With regard to aerobatic aircraft, where structural weight tends to reduce climb and therefore maneuverability, it seems apparent from accident records that inverted load factors in particular should be greater than in present practice. For a relatively small increase in empty weight, negative load factors could be as high as positive design values. The major requirements are identical top and bottom beam cap strips for the new aerobatic monoplanes, heavier compression struts for strut-braced monoplanes, and stronger beam caps and landing wires for biplanes to permit the same load to be carried up or down on the wing structure. Since there already is some form of lower cap strip on every beam, adding to these flanges would mean about 1 percent increase in empty weight for an aerobatic monoplane with cantilever wings (no support struts or wire bracing) with a gross weight of 1200 lb. The wing-load carry-through structure in the fuselage would also have to be strengthened, resulting in a total structural weight increase of about 2 percent in empty weight, or around 17 lb for a cantilever monoplane with an empty weight of 850 lb and a gross weight of 1200 lb. Similar weight increases apply to strut-braced monoplanes and for biplanes.

This seems a reasonable penalty for the additional safety provided. While this margin would not be necessary to cover experienced aerobatic pilots, it is recommended to protect pilots learning aerobatics. After all, practice is necessary to perfect aerobatic technique—and many high load factors are produced along the way.

To further improve structural integrity of aerobatic aircraft, sheet metal construction should be considered because it is more reliable than wooden wings or spars, lends itself to precise stress analysis for maximum design efficiency, and gives more warning of approaching failure than can be expected from a wooden structure. For these reasons, the design of future aerobatic aircraft should probably incorporate aluminum alloy construction, with ±6-g limit load factors throughout the airframe (±9-g ultimate design loading). The slight weight increase anticipated to carry the higher negative g loads could be eliminated by careful use of sheet metal and extrusions combined with competent structural analysis.

While the homebuilt safety record matches that of production aircraft (see Chapter 1), many of these small experimental aircraft are used for aerobatics. The following excerpt [paragraphs 7(g) and (h)] from FAA Advisory Circular, AC, No. 91-48 of June 1977, sounds a note of warning for those who purchase homebuilt plans or completely assembled homebuilt aircraft with the intention of learning aerobatic maneuvers:

Par 7.g. Any discussion of acrobatic aircraft would not be complete without touching on amateur-built aircraft. Many of these aircraft are used for sport acrobatics, competition, and airshow work. It is difficult, however, for a person not knowledgeable of aircraft structures to determine which are acrobatic and which are not. The FAA does not specify structural standards for amateur-built aircraft. Therefore, whether or not the amateur-built aircraft is suitable for acrobatics is largely determined by the builder and the individual FAA inspector who prescribes its operating limitation. Acrobatic maneuvers listed as acceptable for these aircraft are determined during the time the aircraft is undergoing flight testing in a prescribed flight test area. If none are listed, the aircraft is restricted from acrobatic flight. The operator should examine the service history of the aircraft and consult the designer before attempting any acrobatics.

h. Be cautious of any claim that the aircraft is "fully acrobatic." It may never have been designed for that purpose. Acrobatics in amateur-built aircraft should only be conducted in airplanes with service records in that type of operation or which have been designed and built for that purpose.

And even then the design load factors may have been intended only for pilots skilled in aerobatics, the danger being that considerably higher loads may result during "self-taught" attempts to master aerobatic proficiency.

IN THE CABIN   Pilot fatigue takes its toll in mismanaged fuel systems and powerplant controls, missed checklists, sloppy flight technique, misunderstood communications, etc. The major cause of such fatigue originates from the powerplant in the form of noise entering the cabin through the windshield and firewall.

Thicker firewall insulation plus panel deadening materials will reduce engine accessory section and exhaust noise coming through the firewall, while a heavier windshield will effectively reduce propeller and related aerodynamic noise, because the added thickness stiffens the broad expanse of windshield glass and so reduces vibratory noise. Increasing windshield thickness also reduces the likelihood of damage from bird strikes, an occasional and often serious type of accident.

If windshield glass is thickened from $1/8$ to $3/16$ in., the average small airplane will pick up about 5 lb of additional weight plus about \$5 in airframe materials cost (1978 dollars). A similar weight and cost increase also applies to thicker cabin side windows; again the slight increase in weight and cost seems warranted for the additional comfort and safety.

Greater reductions in cabin noise level will require basic powerplant modifications. Locating the engine behind or behind and above the cabin further removes the accompanying propeller, exhaust, and accessory section noise. Thus, reduced fatigue becomes another safety reason for the remote-engine aircraft configuration; the further away a powerplant can be located from the cabin, the quieter and less fatiguing the cabin area. Powerplant noise may also be lowered by installing a more powerful engine than necessary and then operating at reduced rpm to provide required power (as is done in our automobiles). However, this is not a particularly practical solution because a large aircraft engine is more expensive, weighs more, and consumes more fuel even when throttled than a smaller engine of adequate power.

Engine and propeller manufacturers are working to increase efficiencies at lower rpm values for a given power output, but have some way to go before reaching the propeller noise reduction possible by simply decreasing diameter, which brings tip speed rapidly down from the extreme noise-producing sonic-speed range of larger propellers. While some diameter reduction may be realized by using three-blade instead of two-blade propellers, the cost and weight penalties are high. The shrouded fan is an effective step in the right direction (see Chapter 5), and lends itself to remote location as well; so this powerplant package is attractive for our safe aircraft design.

Exhaust mufflers and turbochargers also reduce powerplant noise and should be considered during preliminary design. Their effect is discussed along with other powerplant items in Chapter 5.

FUEL LOCATION   Many powerplant accidents really begin on the ground and could be eliminated by proper preflight inspection. And many preflight checks would be more thorough if the pilot could readily look into the fuel tanks, drain the fuel system conveniently from a single outlet, and easily inspect the oil level and various powerplant components. The basic airplane design determines all of these items, with fuel in low wings being one of the best and safest locations to assure preflight check of the actual quantity aboard. So the low- or shoulder-wing fuel tank positions are more desirable from the safe inspection standpoint than high-wing tanks which require a ladder or gymnastics for proper inspection.

From the structural standpoint, fuel carried in the wings reduces wing bending loads throughout the entire flight range. In addition, this location positions fuel safely away from the cabin in the event of an emergency landing. The real safety of wing fuel tanks is, of course, crash-related; having fuel flying about outside the cabin after impact is at worst more desirable than having it spraying inside. The number of crash-related fires indicates the need for fuel isolation; they also indicate the need to cut off the electrical system prior to or at impact.

Because things may be quite busy in the cabin during an emergency landing situation, consideration should be given to the use of an inertia switch to totally cut off the electrical system at impact. Operating at 3- or 4-g forward load and around 9-g vertical load, such a switch need not be either expensive or heavy —but the added safety would be invaluable.

Considering only those crashes in which aircraft were destroyed, a recent survey found that 25 percent of the single-engine and 41 percent of the multi-engine airplanes burned after impact (Ref. 6.7). Unfortunately, a substantial portion of all such fire-related accidents involve fatalities, so any sound design concept capable of reducing the likelihood of fire after impact should be considered. Remote location of fuel tanks and automatic electrical system cutoff at impact are two such possibilities which should be included in all new aircraft designs.

Another interesting approach is the use of bladder-type fuel cells made from special single-ply flexible material such as Uniroyal No. US764 described in Ref. 6.8. As tested by the FAA, these tank liners successfully resisted crash impacts of up to 27 g forward and 55 g vertical; values certainly capable of retaining fuel well beyond load conditions any passenger might survive. A 60-gal tank made from this material weighs about 18 lb including outlet and drain fittings. This represents a small weight penalty of 0.3 lb/gal for the safety of preventing fuel fires in severe crash conditions.

BEATING MURPHY'S LAW

At the risk of discussing the seemingly obvious: *all parts and components of an airplane should be designed so they can be assembed only in the correct manner.* Murphy's Law is frequently quoted as stating that if something can go wrong it eventually will go wrong. For aircraft assembly operations, we can add two corollaries: (1) If parts can be assembled wrong, they will be—and then are frequently forced to fit or made to operate; and (2) if adjacent moving parts can permit anything to jam and so prevent operation, something eventually will jam.

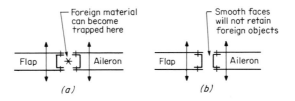

FIGURE 6-16
The effect of control surface detail design upon possible jamming of surface movement. (*a*) Foreign objects can get between flap and aileron to jam operation. (*b*) This design is more difficult to rivet, but it is also more difficult to jam.

By being aware of such potential accidents and by applying a bit more thought during the detail design phase, designers and manufacturers can eliminate many probable accident sources which seem to creep into our aircraft with time. This safety approach also applies to subsequent field modifications added by owners or service mechanics, since operational problems are similar regardless of when a troublesome detail is installed.

By way of examples, consider the following few of many control system possibilities:

(a) Pushrod control systems can jam in flight from trapped water freezing around rod ends and bell cranks at colder altitudes. This can happen during climb, cruise, or descent—and the results can be quite final.

Solution: Use control cables. Pilot effort will break a cable free even if it is encased in a block of ice. (Better yet, prior to flight thoroughly inspect any airplane that has been parked outdoors in a heavy rain or snowstorm, and drain all trapped moisture.)

(b) Adjacent control surfaces should be designed to prevent ice, mud, sticks, stones, etc. from getting between the facing end ribs and jamming operation.

Solution: As shown in Figure 6-16. While (*b*) is somewhat more difficult to assemble, this design is about as impossible to jam as anything could be.

(c) A pilot occasionally attempts a takeoff with external control surface locks in place, similar to attempting to leave a tie-down area with one rope still attached, but the results are much more serious. Of course, this becomes a pilot error accident.

Solution: Design control surface locks to be applied in the cabin and install strong control surface displacement stops at each control surface. Total cost will be about $100 plus 1 lb added to the airframe weight.

(d) Wheel-type aileron control systems require displacement stops on the yoke, but it is possible for these stops to become jammed and so prevent aileron operation.

Solution: As shown in Figure 6-17. The added costs and weight are insignificant, but the added safety could be incalculable.

These are just a sampling of possible detail design safety features, here focused upon the control system; equal effort with accompanying increases in flight safety can be applied to all areas of airframe, systems, and powerplant design.

Some features that invite trouble will never be apparent until the prototype of a new model is assembled. By redesigning to eliminate potential hazard items as they occur, we can beat Murphy's Law. The price is constant attention

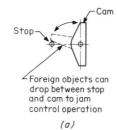 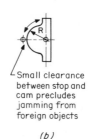

FIGURE 6-17
The effect of control system detail design upon possible jamming of the control system, in this case, the aileron control wheel stop.

to detail design backed by experience, but the benefits of producing a safer airplane include lives saved, reduced insurance costs, and a lasting reputation. Such rewards are obviously well worth the effort required. If any reader cares to pursue this subject at greater length, the disastrous failures recorded in Ref. 6.9 clearly indicate the need to correct design and operational problems when they are first noted.

THE EFFECT OF AIRCRAFT CONFIGURATION

When aviation was a much younger industry, engines were not noted for reliability. For the benefit of safety in emerging commercial transport operations, the use of more than one engine seemed most advisable. In fact, since two were better than one, three were frequently used; through this practice, if one engine had to be shut down, altitude supposedly could be maintained by the two remaining powerplants.

Over the years, improvements in engine, propeller, and accessory reliability represent one of the major advances in general aviation. As a result, twin-engine aircraft no longer really demonstrate superior safety compared to single-engine models. The combination of increased wing loading (high landing speed), fuel consumption, minimum single-engine performance, and lack of adequate pilot technique in twins during single-engine emergencies has conspired to give our modern twins a poorer safety record than the singles, based upon the relative percentage of fatal accidents for each type.

By way of example (Ref. 6.7), considering only small fixed-wing aircraft severely damaged in accidents, 3.85 percent of all active single-engine aircraft and 3.13 percent of the active multiengine fleet were either substantially damaged or destroyed during the survey period. But fatal accidents averaged 13 percent for single-engine vs. 20 percent for multiengine aircraft, which also have a greater burn rate after impact (see page 106). Thus, the twin does not necessarily offer increased operating safety; in fact, it is quite likely that the supposed safety of two engines lures pilots into IFR situations they would avoid with only one engine. The conclusion drawn from current accident data is that a modern single-engine airplane is at least as safe as a twin. The advantages in initial cost, maintenance, insurance, direct operating cost, and pilot qualifications all favor the single-engine airplane. In view of these factors, our new "safety airplane" will have only one engine.

The high- vs. low-wing safety picture is rather hazy; the percentage of accidents totally destroying aircraft and the ratios of accident survivors have been about the same for each type of airplane during recent years. However, the fatalities are slightly greater for low-wing models. This is not particularly surprising because most high-performance singles and twins are of low-wing configuration, and their higher landing (impact) speeds cause the most crash damage. For aircraft of higher wing loading, the advantage of having structure below the floorboards to absorb energy during emergency landings may prove superior to a rugged low-wing transverse beam structure that tends to tear out upon impact, taking the floor and passengers with it. So our safe airplane will have a shoulder wing to permit aft and up visibility as well as provide space for a crash-resistant belly structure. This wing location also offers more ground clearance from obstructions (posts, runway lights, markers, etc.) while permitting more crosswind drift correction during landing flare.

When mentioning energy-absorbing fuselage bottoms, we should again note the inherent safety of flying boat hull structures. Although normal wave impacts do not begin to rival those of an emergency landing in a logged field left in tree stumps, I have seen the results of an amphibian landing wheels up in such an area. All the bottom repair required was straightening two frame lower flanges plus pounding out a few skin dents. Just think how a similar landing with any production landplane could compare in damage; a strong hull structure obviously translates into passenger safety. In addition to its land and water utility, one great advantage of an amphibian is the opportunity to select either land or water for an emergency landing; and if there is no time for a choice, the rugged hull bottom is between yours and the ground.

THE FINAL DESIGN

Reviewing all the options and features considered for a safe airplane, we can combine the desirable items into one configuration such as the preliminary profile of Figure 6-18. While the exact position of the nosewheel or the advan-

FIGURE 6-18
One possible configuration for a stallproof, high-visibility, four-place landplane, with 200-hp Q-Fan-type powerplant.

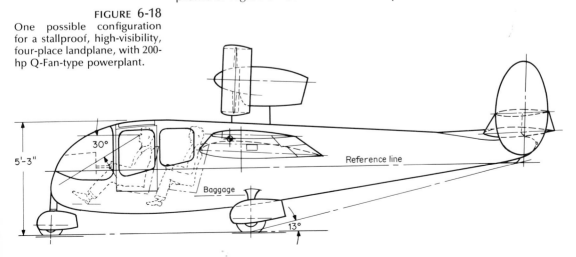

tage of a single vertical tail or even a V tail may be subject to more detailed study, this airplane incorporates the following features intended to reduce opportunities for pilot error accidents and to improve operational safety:

1. Superior visibility forward and down over the nose, to the side, and aft over and under the wing

2. Heavy windshield glass to reduce fatiguing noise and provide increased safety from bird strikes

3. A loading configuration that precludes exceeding aft c.g. design limits

4. Fuselage structure designed for crash load conditions, including shoulder harness for all occupants

5. Tricycle gear designed with sufficient strength to permit crabbed and drift landings

6. Bungee spring nosewheel steering to permit wheel alignment upon runway contact

7. Landing gear and wing incidence arranged to prevent "wheelbarrowing" type nosewheel accidents during high-speed landings

8. Faired landing gear struts and wheels for improved climb and cruising performance

9. Fixed landing gear to eliminate pilot error landing gear retraction accidents

10. Elevator restricted to prevent wing stall at full up-elevator deflection

11. Interconnected aileron and rudder control systems. (Note: Items 10 and 11 result in a stall- and spinproof, two-control airplane.)

12. Elevator and rudder trim tab control systems

13. Lift flaps to reduce landing and takeoff speeds

14. Wing washout plus reduced tip section thickness to retain aileron effectiveness at minimum flight speed

15. A tapered wing to improve roll rate and aileron control response throughout the entire flight regime

16. Wing speed brakes (spoilers) mounted on upper and lower surfaces in sailplane fashion to provide glidepath (approach) control for small field and emergency landing safety

17. Fuel located in left and right wing tanks for safety and to encourage visual preflight inspection of the actual quantity in each tank. Bladder-type fuel cells will be used.

18. An inertia switch to cut off the electrical system at impact

19. A remotely mounted powerplant to reduce cabin noise and to provide increased safety from rotating propeller blades

20. A shrouded fan propulsion system to improve powerplant efficiency and reduce propeller noise

21. Variable-pitch propeller for maximum performance efficiency

22. Thrust line offset to the right to counteract torque effects, reducing the need to apply right rudder during takeoff and climb

23. A color-coded checklist covering pre-takeoff items, best climb speeds, cruising power settings, and landing checks

24. Recommended powerplant control knob distinctive shapes and colors

25. AOA indicator to monitor takeoff climb, cruise, and approach speeds

26. Alternate static source

27. Control surfaces locked in the cabin; movement stops at all control surfaces

28. Last but by no means least: STOL performance

Quite significantly, every one of these items could be incorporated today, since all lie within the range of existing aircraft design technology. The weight increase and estimated cost of many of these recommended features are presented in Figure 6-19 for a reworked version of a typical production airplane, such as a 180-hp Piper Archer or Cessna Cardinal. Note that the estimated weight empty is increased by about 10 percent, accompanied by a probable list price increase of nearly the same amount.

To incorporate these improvements, either the useful load must be reduced, or, more attractively, the structure reviewed to eliminate surplus airframe weight and thereby retain or slightly increase the design useful load. Since weight and cost won't sell, they must be designed out of the airframe and various equipment installations by exercising competent detail design, stress analysis, and weight control.

FUTURE DEVELOPMENT   If we carry the concept of Figure 6-18 a bit further by making the landing gear retractable and adding a single hydrofoil on a retracting strut, it is possible to develop an advanced amphibian design. While the single hydrofoil tends to restrict acceptable center of gravity travel limits, it has the advantage of being less likely to water-loop than a hull or float seaplane when mounted on a smooth bottom surface such as that of our new design.

The hydrofoil proves more stable than a hull at water contact because the hydrofoil support strut has a small load component when yawed (in a skid or drift landing which starts a hull-type water loop) compared to the mass of the airplane moving along above the water surface. Futher, the smooth hull bottom of our design has no keel or chines to dig into the water and cause trouble if

Item	Weight added*		Estimated additional manufacturing cost,* 1980 dollars
	To empty weight, lb	Percent of empty weight	
High-strength tricycle gear	40	2.8	400
More complete soundproofing of firewall and cabin	6	0.4	120
Heavy-glass (³/₁₆) windshield and cabin windows	10	0.7	10
Exhaust muffler	6	0.4	90
Constant-speed, small-diameter, multiblade propeller	15	1.1	$300 more than standard constant-speed propeller
Crash-resistant structure	22	1.5	250
Energy-absorbing foam seats	8	0.6	60
Shoulder harness for all four occupants (instead of only two)	8	0.6	70
Spoilers for glide-path control	10	0.7	250
Crash-resistant fuel cells	20	1.4	250
Total	135	10.2	$1800 direct cost
Safe airplane	1585-lb weight empty		$25,750 new list price with profit and sales markups

* For a 1450 lb empty weight airplane of $23,000 standard list price.

FIGURE 6-19
Estimated weight and cost of adding structural and equipment safety items (based on a 2400-lb-gross-weight, four-place airplane with fixed tricycle landing gear).

yawed. So this spin-free safety configuration can be developed into both a landplane and an advanced amphibian equipped with the hydrofoil described in Ref. 6.10.

$$* \qquad * \qquad *$$

Having established the new arrangement for Figure 6-18, let us now dig into some reasons why previously developed safe designs have not revolutionized the small airplane market. Through this evaluation we can define the criteria for successful introduction of a new generation of safe aircraft; in short, what will sell aircraft safety to the users of personal aircraft?

REFERENCES   6.1   M. H. Waters, T. L. Galloway, C. Rohrbach, and M. G. Mayo, *Shrouded Fan Propulsors for Light Aircraft.* Society of Automotive Engineers Report 730323, April 1973.

6.2   J. W. Turnbow, D. F. Carroll, J. L. Haley, Jr., and S. H. Robertson, *Crash Survival Design Guide.* USAAVLABS Technical Report 70-22, rev. August 1969.

6.3   FAA Federal Air Regulations (FAR) Part 23: Airworthiness Standards: Normal, Utility, and Acrobatic Category Airplanes, paragraph 23. 561 (c) (3) (i). Available from: Superintendent of Documents, U.S. Government Printing Office, Washington, D.C. 20402 Price: $12. Order catalog no. TD 4.6:23.

6.4   FAA Advisory Circular, AC No. 43.13-2A: *Acceptable Methods, Techniques and Practices—Aircraft Alteration,* rev. 1977, pp. 65–75. Available from Superintendent of Documents as above. Price $2.75. Order Stock No. 050-007-00407-9.

6.5   David B. Thurston, *Design for Flying.* McGraw-Hill Book Company, New York, 1978, chap. 14.

6.6   Ibid., chaps. 5 and 7.

6.7   Harold D. Hoekstra and Shung-Chai Huang, *Safety in General Aviation.* Flight Safety Foundation, Inc., Arlington, Va., 1971, pp. 41–51.

6.8   William M. Perella, Jr., *Tests of Crash-Resistant Fuel System for General Aviation Aircraft.* FAA Final Report FAA-RD-78-122, December 1978, National Technical Information Service, Springfield, Va.

6.9   Victor Bignell, Geoff Peters, and Christopher Pym, *Catastrophic Failures.* Open University Press, Walton Hall, Milton Keynes MK7 6AA, England. Available in the United States through: Open University Educational Media, Inc., 110 East 59th Street, New York, N.Y. 10022.

6.10   David B. Thurston, *Design for Flying.* McGraw-Hill Book Company, New York, 1978, pp. 235–237.

7

PAST AND FUTURE SAFE AIRCRAFT DESIGN

Several aircraft configurations satisfying some of our safe design criteria have been developed through the past 40 years, with a few of the designs surviving for some time at relatively low production volume. Let us briefly consider why these prior attempts to introduce stallproof or exceptionally safe slow-flight aircraft have not aroused mass enthusiasm in the marketplace.

PRIOR STALLPROOF AND SLOW-FLIGHT AIRCRAFT

During the period since the late 1930s there have been four certified aircraft designs capable of controlled slow-speed flight and steep descent at low sink rate. In order of introduction these are the Ercoupe, Helio, Maule, and most recently in the United States, the French-designed Rallye series. One of these aircraft, the Ercoupe, was produced in volume as a two-control design which could not be stalled or spun. With the exception of the originally low-priced Ercoupe, these aircraft have at best received limited acceptance. If they are safer and easier to fly, why hasn't every pilot wanted one?

The answer to this question will indicate that safety alone is not enough to create airplane sales, probably because pilots believe their own flight technique provides adequate safety or else they would not fly. However, that is not the actual situation, as even a casual review of detailed accident reports clearly indicates. Every year many highly qualified and experienced pilots become fatally trapped by weather or unexpected situations. And many would survive if their airplane did not stall and could be landed from a controlled steep approach at comparatively low speed. With these premises as a foundation, let us consider the four aircraft, each of which represents a slightly different design approach toward the same goal—controlled low-speed flight.

FIGURE 7-1
The post-World War II Ercoupe and the Aircoupe had the same configuration but different manufacturers.
Howard Levy Photo

THE ERCOUPE   Shown in Figure 7-1, the Ercoupe was introduced prior to World War II as the initial model of a line of aircraft offering exceptional safety; we finally had an airplane that could be flown at low speed without fear of stalling. The original two-control system also offered flight as simple as driving a car. The airplane was stable and could be turned on the ground or in the air by rotating the control wheel just as one would steer an automobile. Although the Ercoupe seated only two in rather cozy proximity, it traveled economically with low power and was relatively inexpensive.

Seriously promoted after the war, over 3500 Ercoupes of both two- and three-control types were delivered. The greatest testimony to the acceptance of this airplane is found in the number maintained in flight status today, owned primarily by sports pilots who wish to fly for minimum investment and usually in VFR conditions.

When postwar production volume saturated the general aviation market, Ercoupe sales virtually terminated along with those of other small aircraft. Subsequent attempts to revive the design, including some modifications intended to modernize the original configuration, were not successful. Probably the major reasons were: (1) the price kept going higher and higher; and (2) this airplane is really too small for today's general aviation market.

The American owner now requires a larger four-place airplane, just as our drivers prefer a large five- or six-passenger car. And while fuel costs may force automobile buyers to accept smaller cars, this probably will not happen with aircraft. A useful airplane must fly at speeds fast enough to save time when encountering headwinds, and do so with sufficient storage volume and lift to carry nearly half a ton of people, equipment, baggage, and gasoline. While it is true that our aircraft should become more streamlined and efficient, a fairly large airplane will still be required to satisfy these practical power and load criteria.

In addition to its size and eventual cost limitations, if flown too slowly, the two-control Ercoupe proved capable of losing considerable altitude before full control recovery was possible. Also, when first introduced this airplane was a handful for many pilots in crosswind conditions because the ailerons and rudders were interconnected, meaning that there was no separate rudder control. In fact, there were no rudder pedals at all—one's feet rested on the floor. Fred Weick, the designer of the Ercoupe, maintained that interconnected controls should not be a problem in crosswinds since the airplane could be safely landed in a crabbed attitude, and many pilots eventually followed this technique.

Unfortunately, there was no substitute procedure for a slip often found necessary during a high approach. Because the two-control Ercoupe could not perform this maneuver, a higher-than-average level of pilot skill was required to land safely on short strips or to set up a steep approach over local obstacles. To make matters worse, as anyone who has flown one of the original two-control Ercoupes will recall, forward vision was virtually nonexistent during a low speed approach.

Both these criticisms would be overcome today through use of speed brakes to control glidepath descent in a normal approach trim attitude. But such glide control was not developed when the Ercoupe was designed, so most of these aircraft were eventually sold as three-control models capable of stalling.

In summation, increased sales price plus inherent handling problems eventually overcame the safety features of the spinproof Ercoupe. Its size and limited capacity were also negative factors. Final market versions of the Ercoupe saw the design offered as a conventional three-control airplane (Aircoupe) of small size and rather high price compared to larger, secondhand aircraft. But when originally offered at a competetive price in the then standard two-place general aviation market, the spinproof Ercoupe outsold all other aircraft by a large margin. So flight safety would apparently command the market—but only if made available at a competetive price and capacity.

THE HELIO   Following a period of design study and flight test, the original Helio airplane was developed after World War II, considerably influenced by the small two-control Koppen Skyfarer. This basic airplane was a stallproof design, in many ways a different configuration from the Ercoupe although offering similar performance and handling characteristics. And it suffered from the same size and approach problems.

As a result, the Helio grew in dimension, became equipped with expensive geared engines, which kept external propeller noise low but increased cabin sound levels, was revised to a three-control system with a single vertical tail, and then incorporated automatic leading edge slats to provide satisfactory low-speed flight characteristics. The resulting payload and handling qualities were simply phenomenal, but so was the price. However, for operation in close areas, off small strips, or out of short ponds on floats, no airplane available compares with the Helio Courier of Figure 7-2.

FIGURE 7-2
The STOL Helio Courier.
Howard Levy Photo

Unfortunately, through various military and commercial modifications, the airplane grew in weight and manufacturing complexity to resemble the World War II German Fieseler Storch. It became a large, expensive airplane that had outstanding STOL and low speed characteristics, but could be stalled and spun. In fact, some of the final models were equipped with stick shakers to avoid stall, because recovery could require considerable altitude. During its lifetime the original excellent and useful Helio Courier had actually evolved into an expensive dinosaur. As a result, most of its outstanding operational functions could be performed equally well and at less cost by the helicopter; so the Helio market slowly shrank to uneconomically low production levels and manufacturing was terminated.

This is a sad scenario, and one which again indicates that although safe flying qualities may command a relatively small and specialized market at a slight cost premium, to realize volume acceptance such aircraft must remain competitive in price, quality, and construction as well as in performance and handling characteristics.

THE MAULE This brings us to an airplane that is competetively priced, offers excellent STOL performance with a good load aboard, and possesses forgiving low-speed handling characteristics: the steel tubing and fabric-covered Maule Rocket shown in Figure 7-3. Developed over the years into a powerful four-place, short-field airplane, the Rocket and related models have not received deserved acceptance. This is primarily because of: (1) modern preference for all-metal, monocoque (sheet metal, frame, and stringer) construction, (2) limited production and sales capability, and (3) rather poor attention to nonessential finish details, particularly in the cabin area. Items 1 and 3 are important to anyone prepared to spend $25,000 to $35,000 for more efficient transportation, as Beech, Cessna, and Piper have learned so well, while limited production and sales capability will always combine to restrict sales. It happens in this case because Maule Aircraft is essentially self-financed and growing at a rate satisfactory to the owners, who prefer to personally manage operations. This is probably just as well since sales demand for these particular airplanes moves along a slow growth path matching the company's financial limitations.

Even though the various Maule aircraft are not stall/spinproof, their slow-speed capability attracts a market that must have this level of STOL performance. To realize the sales growth possible from all-metal construction and better interior finish, the price would have to remain quite competetive with aircraft such as a modern Cessna Hawk XP or a Piper Archer II, while considerable additional working capital and spare parts support would be required.

Although production may seem to be a subject quite separate from aircraft design, the Maule series clearly shows that financial and production capability are also extremely important if a new airplane concept or configuration is to be successfully introduced in some volume—no matter how safe or excellent the handling characteristics may be.

A somewhat similar although smaller airplane, the Piper Super Cub shares a segment of this same STOL market. Backed by Piper sales and production effort, the Super Cub sells at about the same annual volume as the Maule series even though it is a fairly high-priced, two-place, also steel-tube and fabric airplane. In addition, the Super Cub is very capable of stalling and does not possess all the forgiving low speed characteristics of the Maule design. With sales volume of the smaller Piper Super Cub approaching that of the Maule models which have a lower price per seat, we find further proof that financial and production capability exert strong influence upon aircraft sales in a competetive market.

FIGURE 7-3
The four-place Maule Rocket demonstrates a short takeoff.
Howard Levy Photo

FIGURE 7-4
The French Socata Rallye
showing raised horizontal tail
position.
Rallye Aircraft Corporation

THE RALLYE   While not a new airplane, having enjoyed considerable popularity in Europe
for many years, the various Socata Rallye models have been recently reintro-
duced into the United States. Pictured in Figure 7-4, the Rallye departs from
current aerodynamic practice in that the wing is rectangular and fitted with automatic leading edge slats similar to those used on the Helio. Behaving much
the same as a stallproof airplane in its flight handling qualities, the Rallye is a
four-place STOL airplane that is fully controllable at extremely low flight
speeds, but one that can be stalled nonetheless.

Probably capable of filling the performance gap left by termination of Helio
production, the Rallye offers similar capability along with similar sales restrictions; it tends to be rather heavy and expensive for its size. In addition, American buyers are justifiably concerned about the availability of spare or replacement parts for foreign aircraft; the parts situation is usually bad enough from
our domestic sources and much worse from overseas.

Thus, although the Rallye is a very interesting airplane of superior flight
safety and handling characteristics, it remains to be seen whether quantity sales
can be developed in this country considering the current price, useful load, and
spare parts barriers associated with this design; hopefully, they can since this is
a safe and practical STOL airplane produced in a series of models spanning the
range from trainer to high-speed cruise operation.

FUTURE STALLPROOF   The preceding review of four aircraft offering superior low-speed handling
AIRCRAFT CRITERIA   characteristics has underscored the existence of limited market interest in flight
safety at a premium price. As a result of low production, costs of prior stallproof
and slow-flight aircraft tended to escalate right out of buying range for the
safety advantages offered.

To be accepted in volume, it is apparent that aircraft offering stall/spinproof
safety qualities must also be competetive with or superior to the most popular
conventional aircraft in:

1. Handling characteristics throughout the flight spectrum
2. Operating performance
3. Useful load capability (four passengers plus fuel and baggage)
4. Type of construction
5. Ease of maintenance
6. Purchase, operating, and maintenance costs
7. Sales and service support

To this we should add the need for improved visibility in all directions, the flyaway indicator for density altitude and weight variations, the HUD indicator with AOA assistance for takeoff climb and landing approach, a stall warning buzzer/light system to signal marginal control speeds, shoulder harness for all occupants, and a crash-resistant structure in the cabin area. Attention to noise reduction and more efficient operation will also be required to satisfy future market interest as well as government regulations.

All these items represent desirable design requirements, in fact, design specifications, that should apply to the next generation of general aviation aircraft. Most of these features are available and have been for some time. Many should have been incorporated into current production aircraft but have not been for a number of reasons, including production cost, some possible restriction of airplane versatility, age of aircraft designs originally certified many years ago, and the fact that this degree of flight safety is not required by FAA certification specifications.

Obviously, a multipurpose airplane intended for personal transportation, flight training, and aerobatics could not be offered as a stallproof design—but any model intended only for personal aircraft use certainly could be marketed in that version, and should be. However, obtaining maximum safety from the best combination of design features really requires a new family of aircraft. And what major manufacturer is going to develop a new line of aircraft obsoleting all current production models, which are selling well by industry standards?

In addition to the cost and time involved, think about the various ramifications arising from what would amount to an admission that prior designs were not as safe as small aircraft could have been—but now, through a new design concept, would be in the future. Liability suits, dissatisfied customers, reduced value of new and used aircraft inventory, and other related severe strains upon general aviation's limited finances would cover a few of the major problem areas.

If stall/spinproof aircraft are ever to be offered by the leading manufacturers and so have an opportunity to capture their deserved major share of the small airplane market, such development certainly will have to be accomplished through a series of small steps taken one at a time, or by introduction of a completely new configuration developed by a newly organized aircraft company.

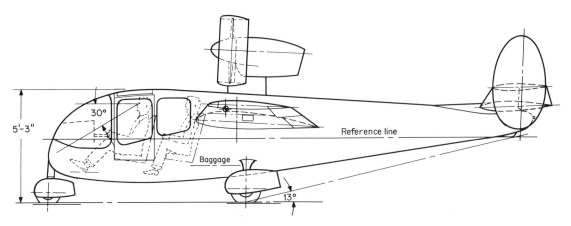

FIGURE 7-5
One possible configuration for a stallproof, high-visibility, four-place landplane, with 200-hp Q-Fan-type powerplant.

On the plus side, FAA certification of any new airplane demonstrating compliance with these safety criteria should be as routine as it ever could be.

The airplane of Figure 6-18, reproduced here for convenience as Figure 7-5, represents only one possible configuration satisfying our specifications. As proposed, this high visibility, quiet airplane cannot be loaded beyond aft c.g. limits. This, of course, means that even if the airplane were severely overloaded, it would not experience elevator control reversal and so takeoff could be safely aborted if the runway is sufficiently long or local terrain flat and free of obstacles.

In addition to the HUD instrumentation with AOA indication, our design has speed brakes (wing spoilers) for approach-path correction. Since the two-control system has no separate rudder action, the speed brakes could be operated by foot pedal for more rapid descent and to kill lift after landing. Fully depressing this floor pedal could also apply main wheel brakes in best automotive fashion.

FIGURE 7-6
Thurston Aeromarine model of the four-place TA16 Trojan amphibian.

Kent Thurston Photo

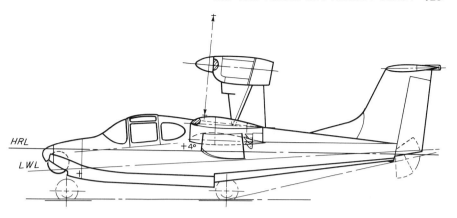

FIGURE 7-7
Profile view of the four-place model TA16 Trojan amphibian.
Thurston Aeromarine

TROJAN AMPHIBIAN

Another raised engine, high-visibility configuration is shown by the Thurston Aeromarine model TA16 Trojan amphibian of Figures 7-6 and 7-7. Note that this all metal airplane has a two-bladed propeller of normal diameter and pylon-mounted engine, plus a T tail to balance offset-thrust-line power changes. Originally designed for EAA homebuilt construction, over 40 of these aircraft are being assembled at this time throughout the United States and Canada. Capable of carrying four people and considerable baggage on long range VFR and IFR flights, the TA16 incorporates several of the safety features described in this book.

RUTAN TWIN

Many other safe airplane variations could be designed, such as the twin-engine Rutan Aircraft model of Figure 7-8. In addition to providing stallproof opera-

FIGURE 7-8
The Rutan Aircraft Defiant, an in-line twin of canard configuration and advanced composite construction.
Dennis Shattuck Photo

FIGURE 7-9
Equator Aircraft P-400 prototype in the foreground with the original test airplane alongside.
Equator Aircraft Photo

tion, the Rutan twin maintains centerline thrust with one engine out and is of composite construction to reduce manufacturing hours and increase operating efficiency.

EQUATOR P-400

The Equator P-400 being developed in West Germany represents another approach offering improved visibility, remote engine location, and very efficient performance. Designed by Günter Poeschel, the P-400 of Figure 7-9 will be offered as a high-speed amphibian of advanced composite construction capable of carrying eight people at 250 mph. With considerable development and test work to be completed, the P-400 will not be available before the mid-1980s.

At present, composite airframe construction is not favorably viewed by stress analysts or the FAA since structural integrity depends upon worker's assembly technique, ambient temperature and humidity, cleanliness of surfaces being joined, and condition of resin or other bonding agent used. However, continued demonstration of equivalent safety and operating life over the next 10 years should pave the way for acceptance of this form of cheaper and corrosion-free construction offering a smooth exterior surface and efficient aircraft performance.

VTOL DESIGN

A different and more advanced design is proposed in Figure 7-10. This profile view of a VTOL (vertical takeoff and landing) airplane has been developed for a test vehicle intended to evaluate handling characteristics. Owing to control and stability problems encountered when hovering, VTOL aircraft require considerable flight development as well as some form of autopilot-induced stability before such a configuration can be considered practical for the average pilot. But the required flight control installation should be quite simple when compared to military fly-by-wire systems, particularly if we accept landings within a field 300 rather than 100 ft square.

With this design approach, the wing can be unflapped and be of the minimum area necessary for cruising flight. Landing speed will not be dependent upon wing lift, so large wing area and high lift coefficients are not required. But engine thrust is critical because thrust must equal or exceed gross weight to provide vertical lift, and this thrust must be delivered at the most critical density altitude condition anticipated for service operation.

As a result, propeller design and power transmission become major problems along with stability. For economical production and practical maintenance, the propellers must be of the variable-pitch, high-thrust, rigid-hub type used on high-performance aircraft, instead of the flapping rotor used on helicopters. At the present stage of design, rather than using a mechanical drive system, power would be transmitted through piping from a large hydraulic pump mounted at the engine to a hydraulic motor mounted on each wing panel. These motors in turn rotate the two lift/propulsion propellers in opposite directions and are interconnected for fail-safe operation.

By the time a four-place VTOL airplane is developed during the late 1980s or early 1990s, we should have relatively cheap turbines available through use of ceramic blades, vanes, and other critical components. One of these light powerplants could then be mounted on each wing for direct power application, eliminating the weight and cost of any drive system other than a safety interconnection between engines.

Although this design has not been flown to date, the VTOL concept offers the most safety and utility of any aircraft configuration. In fact, it is possible to

FIGURE 7-10
Profile view of proposed VTOL test vehicle capable of operation from land or water.
Thurston Aeromarine

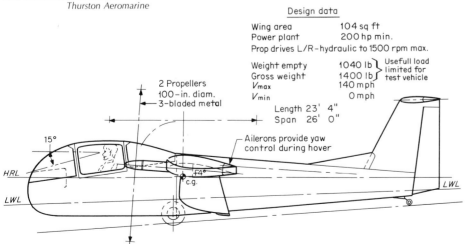

Design data

Wing area	104 sq ft
Power plant	200 hp min.
Prop drives L/R	hydraulic to 1500 rpm max.

Weight empty	1040 lb	Usefull load limited for test vehicle
Gross weight	1400 lb	
V_{max}	140 mph	
V_{min}	0 mph	

Length 23' 4"
Span 26' 0"

2 Propellers
100-in. diam.
3-bladed metal

Ailerons provide yaw
control during hover

15°

HRL

LWL

f4°
c.g.

LWL

LWL

develop one basic airplane capable of landing on land, water, ice, and snow. Sometime between now and the end of this century, the airplane and helicopter should be successfully joined to provide an easily handled, rigid-rotor (propeller) driven vehicle capable of vertical flight, economical cruise, and zero forward speed at lower altitudes. With these improved levels of utility and safety, public acceptance should increase VTOL production to such a degree that all this could be available at today's prices.

As proposed, this machine would combine practically all functions presently expected of different aircraft into one airplane design, thereby providing a major advance in aircraft development. With the speed-range capability offered, we will finally be able to defeat weather on its own ground; when things become too murky, simply rotate the propellers to provide vertical lift and gently settle to the earth in an open area. The autopilot stability system plus the high lift propellers will have landed you safely until weather conditions improve.

WHAT PRICE RESEARCH?

Development of any one of these advanced types of safe aircraft will require considerable amounts of money, something normally in short supply in the general aviation industry. In fact, once a configuration has been carried along the experimental trail far enough to provide satisfactory control and handling characteristics, about an equal sum is necessary to demonstrate compliance with FAA certification requirements. So the price to carry a radically new design from sketch to certification can be enormous by general aviation standards.

In view of the potential growth possible through introduction of aircraft possessing truly safe flight characteristics, it is difficult to understand why the FAA and NASA do not allocate adequate funds for small aircraft development. Our foreign trade balance would be even worse than at present but for the large export business enjoyed by the major general aviation and commercial transport manufacturers. Since we reached this favorable recognition in great part because of research conducted by the National Advisory Committee for Aeronautics (NACA) prior to World War II, the time seems right for another research push to develop the next and future generations of safe and efficient personal and business aircraft.

Repeated attempts to determine the approximate amounts of government support recently allocated to small aircraft design, construction, handling qualities, and flight instrument research have been unsuccessful, principally because there has been very little funding. NASA has developed a few new airfoils having low drag and excellent stall characteristics, but these have large section moments producing high balancing loads, which means they are inefficient wing sections. Low moment airfoils of low-drag profile already exist; combining these with elevator restriction to prevent stalling will provide an efficient stall/spin-free airplane when flown with a two-control system. Thus, with proper airplane design, the new NASA airfoils have little to offer. This indicates that any productive research program must be fully planned before work can begin on an item-by-item basis, and selection of the various research projects should be influenced by useful input from general aviation manufacturers and operators.

Industry apparently will not or cannot fund development of basically new configurations or systems offering the degree of safety necessary for broader acceptance of the personal airplane. This means that the NTSB and the FAA are basically scorekeepers of accident statistics, with little power to favorably influence future design.

Unfortunately, space research has virtually no application to small aircraft except in the field of more reliable avionics. What's needed is active research and development support leading to new aircraft and flight systems; NACA originally performed this service very well and NASA should do so now. The gains possible in lives saved, more efficient transportation, and improved trade balance should be an adequate goal for responsible government support—and now is the time to begin. Our future small aircraft will be most exciting if we can ever get them off the ground.

$$\ast \qquad \ast \qquad \ast$$

Although this chapter ends our direct review of design details intended to reduce opportunities for pilot error, I'm sure you have noticed that many—if not most—of these items are closely related to operating procedures or flight technique. In fact, *design and flight requirements are quite inseparable.*

In view of this, let us see what might be done to make flight operation safer by eliminating a few confusing procedures and dangerous obstructions as well as through the adoption of some recommended communication, flight technique, and maintenance improvements.

The balance of the book relates directly to flight safety, beginning with an analysis of the various types of accidents that occur along the flight profile. Curiously enough, as we shall see in the following chapter, the accident rate steadily increases as flight continues from takeoff toward final approach and landing. Many of these accidents are caused by pilot error, indicating the accident rate can be lowered by a combination of improved aircraft design and flight procedures.

PART THREE
FLIGHT SAFETY

ALONG THE FLIGHT PROFILE

Knowing where most accidents occur during flight will assist in defining the various operational problem areas. Unfortunately, a large percentage of flying accidents are the direct result of weather conditions or some situation induced by weather. Since we can do little to counteract nature at its worst other than obtain a thorough preflight briefing or execute a 180° turn in time for safety, a certain percentage of pilot error weather accidents will always remain with us as a legacy of unpredicted and unpredictable weather.

This fact of life was recently highlighted by a personal experience that rapidly terminated a holiday trip. We had just taken off and were climbing out to the south when suddenly everything went white—the wings, the cowling, our view out the windows—snow. It shouldn't have happened, but it did. Much as fog frequently forms without warning, blocking out everything along a cool coastline, a local snow squall had suddenly become an area snowstorm.

We had left our home airport on the edge of the Alleghenys after checking destination weather by phone and finding clear skies and brilliant sunshine ahead. Our local flight service reported some "lake effect" snow showers to the north which were expected to move out shortly, and 25-mi visibility. There had been one ominous note, however; when calling for takeoff, the tower directed my attention to a local squall northwest of the field. Apparently, that was the beginning of the storm that soon surrounded us, but we were heading southeast and supposedly into clear air.

To make a long story short, I filed IFR once it became obvious we were in more than a local snow cloud, then located our position by taking a radial from the omni and an ADF bearing to the approach beacon. We were in no real trouble, although a bit of rime ice was building along the wing leading edges (and on the stabilator, too, as I learned on landing). No one panicked, but my passengers were mighty silent and sat very still.

Our field was located in mountainous country where 3600 ft above sea level is about the lowest safe altitude with some margin in blind conditions—and we were blind. Although the tower immediately cleared us for the approach as number one to land, I advised that our altitude would be maintained until we reached the beacon inbound on our instrument approach. (Unfortunately, about 2 weeks later an airline crew headed for Dulles International Airport did not follow this same procedure; instead they descended when cleared for the approach and flew into a hillside.)

We landed safely, having broken out into the clear below the storm when inbound on final, and I finished our journey to nowhere with one of the best landings I have ever made. I don't know whether this occurred because I was keenly alert, or because the Warrior handled better with rime ice on the wing and stabilator leading edges.

More to the point—if I had not been instrument rated, I do not know whether I would have taken off after the tower mentioned the snow shower northwest of the field; however, I believe we would have returned to the hangar. I do know we would have been part of the year's accident statistics if I had not been IFR rated and current in local procedures—and all because of a sudden storm that was not forecast and could not have been expected from local observation. So even with flight planning and preparation, weather would have caused another accident except for the protection of instrument training. In addition to sudden changes in predicted conditions, unexpected flying weather sometimes arises from inadequate planning or from "going just a bit further to see if things will improve." These quirks of human nature continually contribute to primary accident statistics in the form of pilot error, with weather listed as a secondary cause.

The combination of airplane, pilot, and weather produces a flight accident profile quite regularly approaching the distribution of Figure 8-1. While these data were based upon reports for 1975, the values are similar to those for 1968 and 1969 of Ref. 8.1, as well as those of more recent years.

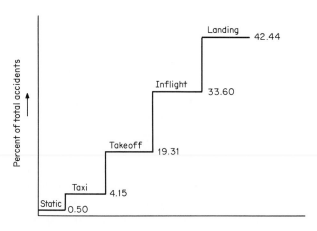

FIGURE 8-1
Accident frequency vs. flight profile. Percentages shown for 1975 remain remarkably consistent from year to year.

Note that the percentage of accidents increases for each phase of progress along the flight profile, beginning with a small value from the moment the airplane is occupied. In fact, there is considerable evidence that many accidents begin even before that as the result of poor or no weather briefing or preflight inspection. While landing requires the most skill of any phase of flying, and so represents about half the accidents reported, many landing accidents actually stem from problems developed in flight. Since these are caused by powerplant, weather, radio, or aircraft systems and equipment among other possibilities, let us quickly review what can and does happen and with what frequency. This information determines the major flight problem areas for subsequent discussion and improvement.

STARTING OUT   There is considerable hangar or tie-down damage, frequently referred to as "hangar rash," that remains unreported or unnoticed until an accident occurs or unexpected handling responses are experienced in flight. Locally bent or flattened leading edges of wings and tail surfaces fall into this classification, along with broken antennas and cut tires—all good reasons for a thorough preflight inspection prior to getting aboard.

From the moment the starter key is turned, accidents happen. And they may occur before a pilot is even aware the key is in the airplane. Turning the engine by hand during preflight inspection is sound practice if the airplane has been idle for a half day or longer—but only when the ignition key is *out* of the starter switch. More than one person has been surprised by having the engine start during preflight "propping" to clear the cylinders. (*Propping* or *pulling through* is turning the engine over by hand-rotating the propeller prior to starting.)

A corollary condition also exists: while it is illegal to start an engine by cranking the propeller unless someone at least qualified to handle the engine controls is in the cockpit (or cabin), each year a few people decide to do it themselves. Although an airplane is supposed to have the main wheels chocked during hand-starting, I presume most of us have seen the safety training film of an actual situation in which an unpiloted airplane chased its owner around and around an airport before finally coming to rest midway through other parked aircraft.

When an engine has been standing for awhile and has not been pulled through by hand prior to starting, an accident may result if fuel or oil has collected in the cylinders. When the engine fires, trapped fluids can cause bent piston rods, cracked cylinders, or a damaged crankshaft. While such accidents may not cause bodily injury, they surely can harm one's budget and can be avoided by proper preflight procedures. It is also possible for damage of this type to remain hidden until something fails in flight.

And even when all preflight steps are properly taken, overpriming an engine in cold weather can result in excess fuel collecting within the cowling. If a backfire occurs, an engine fire may result, and does a few times each year.

Once the propeller starts turning, people do walk or reach into it, though fortunately, very infrequently. Probably one hidden advantage of security fences at controlled airports has been keeping people away from ramp areas and whirling propellers. There are not many other things that can happen to an airplane when standing still, so the static accident rate is quite low, but once motion begins the opportunities increase very rapidly.

TAXIING   While taxiing is generally considered to be the trip from a ramp or tie-down area to the takeoff point, accident statistics also include ground motion following landing rollout as a taxi operation. Surprisingly enough, about twice as many accidents causing substantial damage regularly occur while an airplane is taxiing back after a landing as happen during travel to the active runway for takeoff. Possibly, pilots are more relaxed or less vigilant following a landing; and when off guard, accidents follow. Although the reason for this difference is not readily defined, taxi injuries are rarely serious in either case. But while the airplane is moving on the surface, the causes and types of accidents are varied and numerous if not fatal.

Tackling weather first, we find taxiways becoming tests of skill because of high crosswinds or poor visibility. While takeoff in zero-zero visibility and ceiling is a very suspect and frequently fatal procedure, pilots and their aircraft continue to disappear into the fog—only to suddenly stop with a bang after taxiing into a ditch, post, building, other airplane, embankment, etc.

Strong crosswinds cause brakes to fade if taxiing in a tailwheel or nonsteerable nosewheel airplane. When this happens, the plane gracefully weathercocks in slow, ground loop fashion and heads for the same ditches, posts, buildings, etc. Strong gusts have a similar effect on aircraft with steerable nosewheels, so the amount of wind you and your airplane can handle is always an important consideration during ground operation.

Even in clear weather, experienced pilots in tailwheel-type aircraft have such poor visibility forward they occasionally taxi into parked or moving aircraft, potholes, ditches, etc., while pilots used to steerable nosewheels lose control and cause collisions when operating free-swiveling nosewheel aircraft.

Water and snow conditions require special pilot training frequently found conspicuous by its absence when the hearing board convenes after the accident. So we find seaplanes taxiing in sea states exceeding airplane and pilot capability. When an airplane is on floats, high winds cause capsizing or even overturning end for end if the controls are not properly managed or if the seas become too high.

And snow frequently covers ice ridges that can shear off the nosewheel strut of a fast taxiing airplane, while icy or snowy surfaces send aircraft skidding into snowbanks, runway lights, posts, buildings, etc.

Poor maintenance of airport taxi and runway areas with accompanying soft spots, potholes, and loose surface fragments cause aircraft to shear off landing gear legs, damage propellers, or even turn over if one or both main wheels drop below the surface. Quite obviously, many of these accidents would be averted if the pilot had better ground visibility. When proper design could improve visibility forward and so reduce the likelihood of accidents of this type, it hardly seems fair to blame most of them on pilot error. However, that is the generally stated cause—and we are not even off the ground as yet.

There are a few other surface accident situations that fall into dubious areas of responsibility. For example, brakes may fail during taxi with resulting loss of control and subsequent collision damage. Is this the result of poor maintenance by the owner/pilot, and so a pilot error accident; or is the brake design faulty, causing a design liability accident? Or, when a pilot taxis into heavy wake turbulence spreading across the airport, is this pilot error, tower responsibility, or the fault of the airline whose transport just landed? Many similar ground accidents occur; since their cause has effect upon personal safety, traffic control procedures, liability insurance costs, and aircraft insurance premiums, their prevention is important to all involved.

TAKING OFF

Taking off should be a simple procedure of checking magnetos and carburetor heat, controls operation, instruments, and propeller pitch response; setting gyros and altimeter; making sure there is no one on final even when cleared by the tower; lining up the runway; and pushing the throttle home. And it usually is. But occasionally the problems caused by overloading, exceeding c.g. limits, powerplant failure, density altitude, weather, sea state, or local terrain cloud the operation while adding to accident statistics.

With the exception of landings or cruising flight, the greatest number of accidents of a single type occur during the takeoff climb. This is also a high-fatality-type accident, exceeding the most fatal landing condition (which is final approach VFR) by about two to one. Quite often, when the takeoff climb begins, a number of the unfavorable conditions listed above combine to create an unmanageable airplane.

It is not unusual to find an airplane both severely overloaded and having the critical center of gravity located outside approved flight limits. If we add a hot day and throw in a high-altitude airport, the density altitude effect literally becomes the capstone of a poorly planned flight. Since this combination frequently occurs, representing a most basic form of irresponsible pilot error, what can be done to improve the situation? Analyzing the cause of such accidents usually reveals poorly maintained weight and balance logs, a pilot unaware of or not concerned about allowable gross weight and c.g. limits, a pilot unaware of density altitude effects, or combinations of these items.

Improperly maintained weight logs or lack of proper concern for weight limitations are violations of FAA regulations, to say nothing of showing little concern for human life. But if the correct weight and balance data are not available to a conscientious pilot, the best preflight calculations may be futile and deadly dangerous. Many aircraft are extremely sensitive and may experience *control reversal* if loaded too far behind aft c.g. limits. (To explain control reversal: a pull on the elevator control to begin climbing suddenly changes, as speed increases, into a push required to maintain the climb path.) This condition can be fatal at low altitude if the airplane stalls before the pilot regains control, particularly if trees, poles, or buildings are located along the climb path—and they usually are.

It is possible to design aircraft that cannot be loaded beyond approved aft c.g. limits as shown in Chapter 6, but only common sense and good arithmetic can preclude overloading. The forward c.g. is not so critical, although weight too far forward will make approach trim more difficult and may prevent seaplanes from taking off; however, there will not be any damaging reversal of control force during flight.

Density altitude effect occurs anytime ambient temperature exceeds 60°F or the takeoff surface is above sea level—so density altitude is always with us to some degree but does not require attention unless altitude exceeds 1000 ft or temperatures rise above 75°F. Density altitude affects takeoff speed relative to the runway since lift is dependent upon mass flow of air; when air is less dense, velocity over the ground must increase to provide the required lift. This means a longer run to reach takeoff speed; and when combined with overloading, the flight may end in a forced ground or water loop in order to miss the approaching trees or shoreline, or by collision with any obstructions bordering the takeoff area.

As an alternative, there is a great instinctive reaction to pull back on the elevator control as the end of the runway grows ever nearer, producing at best an even longer takeoff run or possibly a forced takeoff below flight speed, resulting in a subsequent mush-in hard landing or stall/spin finale. Since this condition is amplified by density altitude effect, pilot attention to indicated airspeed or possibly a takeoff speed indicator seem to be the only solutions to this stall problem. Very significantly, approximately 25 percent of all stall/spin accidents occur during the takeoff phase of flight; they are mainly caused by attempting flight before reaching flying speed.

There are other reasons for ending takeoff just beyond the airport boundary; principal among these are powerplant failure and local weather. Many powerplant problems are the result of improper preflight inspection. The engine will not run without fuel and may run out of oil on a long trip; both levels should be checked prior to flight, but accidents due to these causes repeatedly happen. Or, takeoff may be attempted with the fuel selector shut off or placed on an empty tank with just enough fuel remaining in the carburetor to get a few feet into the air before the crash. Proper checklist procedure should prevent this happening.

With jet fuel now available at many airports, it is possible to receive a tank of kerosene instead of 80- or 100-octane fuel; however, our reciprocating engines won't run on heavier fuels because of differences in carburetion and compression ratios, so again we have a takeoff terminating a few feet after becoming airborne. Improper or inadvertent use of the mixture control or forgetting to enrich mixture when descending from altitude can also terminate flight very rapidly when at low altitude.

By way of condensing a long list of additional powerplant accident causes, it is sufficient to note that oil line, fuel line, and exhaust system failures can end a flight just as surely as an engine compartment fire, a broken propeller, a ruptured fuel tank, or severe induction system icing. And occasionally an engine crankcase, gear box, or other major engine component or accessory failure occurs during takeoff, resulting in forced landings for a single-engine airplane, or a treacherous limping turn back to the airport for a twin. With all these possibilities, it is easy to see why powerplant failures constitute such a large percentage of first type of accident as shown on Figure 1-3.

Nature also takes a frequent toll during the takeoff. Gusty crosswind conditions cause loss of control, ending up in ground or water loops, collisions with objects near the runway, and capsizing on water. Wind shear and thunderstorm turbulence can destroy aircraft taking off in conditions described in Chapter 9, while wake turbulence is becoming a steadily increasing problem as covered in Chapter 12.

Trees, buildings, and other upwind obstructions create severe local ground turbulence resulting in flow reversals, downdrafts, roll turbulence, and gusty air causing loss of control, stall at takeoff, or inability to climb faster than the hillside rises ahead.

Heavy sea states caused by extreme wind conditions damage hull and float bottoms, pitch seaplanes out of the water for a stalled crash, and cause capsizing during the takeoff run.

While the zero-zero takeoff is a marginal IFR procedure for the average pilot, every year a few VFR pilots try their skill with limited or no success. Personally speaking, unless there is sufficient ceiling to permit a safe turn back to the airport (300 ft or more above ground as discussed in Chapter 10), conditions are not safe for taking off on an IFR flight—to say nothing of a zero-zero takeoff by a VFR pilot.

Fog and rain also claim their share of VFR pilots flying off into ceilings lower than expected, particularly when encountered at night without any reference horizon available from runway lights or illuminated city streets.

The list could go on, but with the items covered so far it is obvious that many trained decisions are necessary for every takeoff, quite a few of which could be replaced by design features possible today, thereby removing numerous opportunities for pilot error right from the time flight begins.

IN FLIGHT Although initial climb-out is statistically considered part of the takeoff phase of flight, accidents occurring during climb to cruising altitude are listed with in-flight operation. Again we find stall/spin accidents resulting from loss of flight speed during the climb, presumably due to pilot inattention, turbulence, or weather, with weather conditions taking a steady toll all along the way.

Normal cruising flight accidents of various types represent the largest in-flight group, with additional totals added primarily from buzzing, low passes which proved too low, agricultural operations, and aerobatics. These are predominantly stall/spin types and so are heavily fatal, with aerobatics clearly in the lead with about 75 percent usually fatal (28 deaths from 37 aerobatic accidents during 1977 is typical).

Weather accidents during cruise result from improper or no preflight preparation, suddenly encountering unpredicted weather such as imbedded thunderstorms or severe snowstorms with high turbulence and gusts, flying into mountains shrouded in fog, loss of control due to IFR conditions with VFR capability, disorientation in instrument conditions, and severe icing.

Ice, one of the worst and most elusive weather villains, can block the engine induction system, reducing or eliminating power while adding to airplane weight through buildup on surfaces and fuselage. In fact, ice is probably one of the major causes of unknown accidents; if it results in powerplant failure with a subsequent fatal forced landing, the evidence may melt in best detective-story fashion before a rescue party can reach the crash site. As noted in Chapter 5, all powerplant systems should be equipped with alternate automatic induction air inlets to help prevent such accidents.

In addition to providing airspace when going from A to B, normal cruising altitudes are, of course, also part of the realm of practice flying. Various types of accidents, including midair collisions, occur while proficiency is being honed to a fine degree.

As we leave cruising altitudes, we find that about as many accidents occur during descent as during climb, and with the same degree of fatality. A large percentage of these represent collisions with the earth caused by weather, and quite a few are again the result of low airspeed with a stall/spin ending.

The final large in-flight group is classified as "other causes," which refer to search and rescue, coyote hunting, cattle roundups, sickness, unknown causes, etc. This is also a stall/spin category, indicating loss of flying speed, with an attendant 60 percent fatality rate (Refs. 8.2 and 8.3).

LANDING The accident frequency goes up as airplanes come down. Returning to the surface causes by far the largest percentage of all accidents during the landing phase as shown by Figure 8-1. The number one total of any single type occurs in the moments of leveling off and touching down; closely followed in second place by landing rollout accidents prior to taxiing in from the runway. Even though these represent about 28 to 30 percent of all accidents each year, they are rarely fatal, but do result in a considerable number of substantially damaged aircraft, thereby increasing hull insurance costs while causing many embarrassing explanations.

The third major cause of landing accidents develops during VFR final approach, and is again heavily laced with failure to maintain flying speed resulting in stall/spin accidents. So we find that stall/spin problems follow every portion of the flight profile from the moment of takeoff until the final landing phase.

Traffic around airports due to normal pattern circling, VFR go-arounds (touch-and-go landings, etc.), IFR final approaches, and IFR missed approaches are the principal causes for the remaining types of landing accidents. A few of these are midair collisions, a large number again are stall/spin endings (Ref. 8.2), and some are overshooting/undershooting accidents.

Powerplant system failures once more cause a steady stream of problems during the circling and approach phases of landing, partially because trouble developed elsewhere and a landing is being attempted at reduced or no power, partially because of mechanical failure, and partially the result of fuel or engine control system mismanagement.

Fortunately, it is possible to redesign some of the fuel and engine monitoring and control procedures to reduce the opportunities for such accidents. However, there is not much to be done for the pilot who takes off with little or no fuel or who habitually tries to extend the flight range right up to dry tanks —except to increase that person's insurance premium rather than spread the loss over us all.

Weather, of course, again raises its head, particularly in IFR accidents, but airport location must share much of the blame as will be noted in Chapter 12. And, in conclusion, the accident rates for single vs. twins and high- vs. low-wing aircraft remain about the same percentages for each type, with twins having a slightly higher ratio of powerplant and weather-induced serious accidents (Ref. 8.1).

$$* \qquad * \qquad *$$

These different accidents have not been discussed to discourage flying, but rather to clearly underscore the continuing loss of life usually credited to pilot error. When we step back and summarize this chapter, it becomes evident that every year weather, stalls, powerplant system operation, and airport conditions are the real causes of a great many so-called pilot error accidents. For example, consider the number of wrecks continually occurring from loss of flying speed with a subsequent stall/spin ending the flight. When there are 11 ways to either warn of impending stalls or to prevent them entirely as listed in Chapter 3, doesn't it seem about time the flying public was offered at least the option to choose protection from stalls?

While aerobatic aircraft must stall to perform complex maneuvers, we do not require that ability for normal pleasure and business flying. If fighter aircraft flown by some of our most capable pilots employ multiple-control support systems to prohibit stalling (which can be catastrophic at high speeds or altitudes), why isn't the same degree of safety applied to our small aircraft?

Unfortunately, similar questions can be raised about other flight accidents frequently blamed upon pilot error, as the following chapters will clearly show.

REFERENCES   8.1   Harold D. Hoekstra and Shung-Chai Huang, *Safety in General Aviation*. Flight Safety Foundation, Inc., Arlington, Va., 1971, pp. 34, 47–59.

8.2   *General Aviation Stall/Spin Accidents 1967–1969*. National Transportation Safety Board Report NTSB-AAS-72-8, September 1972, p. 11.

8.3   *Annual Review of Aircraft Accident Data—U.S. General Aviation Calendar Year 1975*. National Transportation Safety Board Report NTSB-ARG-77-1, January 1977, p. 26.

THE HARD EARTH

HARD LANDINGS The NTSB defines a *hard landing accident* as one caused by *stalling* onto or flying into a runway or other intended landing area. By way of further definition, glassy water is an excellent example of an intended landing area, and one responsible for an increasing number of seaplane accidents.

In a very real sense a hard landing is a form of collision with the earth, but is differentiated from a collision accident, such as flying into a hill, by being directly associated with an intended landing. The distinction in type of accident between a collision and a hard landing would seem to be drawing a fine line, but one apparently found useful in separating landing accidents from other types of collisions with the surface.

At any rate, we have about 250 to 300 hard landings recorded every year as a first type of accident. Although the hard landing percentage of total accidents has dropped from nearly 13 in 1948 to around 7 percent in recent summaries, the fact that the word "stalling" appears in the hard landing definition raises another flag signaling the need for improvement similar to the stall/spin prevention measures reviewed in Chapter 3.

The reduction in hard landing accidents between 1948 and recent years is probably due to the introduction of tricycle landing gear with its superior handling characteristics, the use of stronger landing gear and support structures, and increased flight training in stall recognition. But even so, nearly 60 percent of all landing gears involved in hard landing accidents collapse in the process, while the airplanes proceed to ground-loop, nose over or roll over, and collide with poles, trees, fences, ditches, embankments, etc. Single- and twin-engine aircraft seem to experience nearly the same percentage of hard landing accidents, so recovery does not materially improve with two engines once a stall or high bounce occurs during landing.

Every year some aircraft succeed in hitting so hard that secondary accidents are listed as resulting from a stalled or mushing impact following a high bounce after initial ground contact. Again, after the first type of accident some second-ary accident of magnitude usually follows to compound the event.

In fact, we can state for all aircraft accidents that: *an initial accident will be followed by one or more secondary accidents that may be of greater magnitude than the initial event.* As examples: a hard landing (first type of accident) resulted in collapsing the landing gear (second type of accident); or engine failure at altitude (first type of accident) resulted in a crash landing (second type of accident), which injured a passenger (third-level event).

Hard landings are particularly prone to end up with a related accident because frantic efforts are often made to recover or correct the situation once a low-altitude stall or high bounce is underway. Engine torque, improper pilot response or technique, crosswind gusts, and other factors may cause the airplane to head in almost any direction out of control—including a ground loop or roll-over accident.

The action of sharp gusts during landing is one situation that cannot be controlled. Rapid changes in wind velocity can stall any wing if flight speed is permitted to get too low during the approach. So the ball comes back to flight technique as the final and controlling factor during landing, especially in crosswinds or gusty weather. An AOA indicator plus approaching at an airspeed increased by an amount equal to gust intensity variations would safely overcome these conditions, and should be considered essential for reducing hard landing accidents.

TOO LITTLE OR TOO MUCH

Occurring with a frequency similar to hard landings, under- and overshooting landings consistently account for some 300 to 360 accidents per year. As Figure 1-3 indicates, overshooting normally leads undershooting by about 2 to 1, but 1977 closed the gap with 161 overshoots to 157 undershoots for a total of 318. Nonetheless, overshooting a landing remains more popular than landing short.

Figure 9-1 depicts the accepted definition for these types of accident, which can be more fully described as follows:

An *overshoot* is caused by landing too fast or too far down the runway or

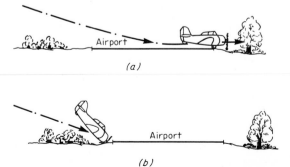

FIGURE 9-1 Undershooting or overshooting a landing usually results in a secondary collision or a ground loop type of accident. (a) Overshooting a landing results from landing too fast or too far down the runway. (b) Undershooting a landing means making surface contact short of the runway.

other intended landing area, resulting in: (1) running off the end of the landing area (including resulting collisions); (2) ground-looping, nosing down, or over-turning off the intended landing area; and (3) landing beyond the intended area.

Collisions resulting from overshooting would be listed as a secondary type accident, and in addition to item 2 above may include: collapsing the landing gear; controlled collisions with the surface; collisions with wires, poles, trees, fences, buildings, ditches, embankments, parked aircraft, runway lights, and approach lights; stalling or mushing in; and complete airframe failures. Quite frequently the end result is very thorough damage.

An *undershoot* results from landing or making contact with the surface or other object short of the runway or other intended landing area.

Quite significantly, there are few reports of over- or undershooting accidents listing aircraft damage of minor nature; the airplane is usually described as sub-stantially damaged or destroyed, although the fatality level is fortunately low. But fatalities from undershooting consistently run 2 to 1 higher than from over-shooting, so if you must choose one or the other, prefer overshooting—the sta-tistics apparently favor that decision.

The accident distribution between student and commercial pilots is surpris-ingly similar for under- and overshooting, with private pilots clearly leading the field about 4 to 1 as shown by Ref. 9.1. Airline pilots also have approach prob-lems as reviewed in Ref. 9.2, which includes some useful IFR procedures. In order to develop methods for reducing these types of accidents, let us review the basic causes for under- and overshooting.

Crashes from under- and overshooting result from: (1) pilot error in selecting the approach speed and configuration for existing conditions; (2) pilot error in not correcting the approach glide path in time; (3) pilot error in not deciding early enough to go around again and begin the approach anew; (4) pilot error in improper application of rudder with sudden use of full power; and (5) local weather conditions. And even weather as a cause for this type of accident can often be traced back to pilot judgment. Fortunately, much can be done to lower the frequency of these accidents by combining proper stall recognition training and missed approach procedures with the previously mentioned AOA indi-cator.

PILOT ERROR   Undershooting is often the result of attempting to stretch an approach until it winds up with a stall/spin/mush ending when, in fact, power should have been applied to reach the field or to go around again for another try. So providing ample stall warning or a stallproof airplane would eliminate a large percentage of such accidents. The design features outlined in Chapter 3 apply here, but even a stallproof airplane can end up short of the runway if it is below the proper approach path to begin with. While the resulting accident and damage would be less severe than for an airplane spun in on final, there is no real substitute for pilot technique and a properly executed approach. And it is the stall/spin finale which makes current undershooting accidents more fatal than overshoots.

Overshooting due to pilot judgment may be the result of: arriving at a field sooner (and therefore higher) than expected; last-minute orders from the tower; trying to miss other aircraft in the pattern or on final; previously encountering much higher headwinds at altitude, carrying over into assumption that these velocities also exist near the ground; poor flying technique such as approaching too fast or downwind; and poor communications with the tower or unicom regarding the active runway or local traffic.

There really is very little that can be done designwise to reduce this type of accident; certainly, stall warning or special instrumentation would be of no use in most of these circumstances, which to a large extent explains why overshooting outpaces undershooting.

Additional flight training or flying more frequently would probably help, but there will continually be situations where for some reason pilots are too high and still decide to see it through rather than go around. Making a second approach is no disgrace—carrier pilots do it all the time—but landing in the trees at the end of the field can be mighty embarrassing and equally expensive.

With many pilots renting aircraft and also being approved for flight in several different models, the tendency to vary airspeed indicators between miles per hour and knots from airplane to airplane presents an additional source of under- and overshooting—particularly for relatively inexperienced pilots. If a pilot is trained in mph approaches, the same airspeed indication in knots means a fast approach and overshooting a small airstrip. If a pilot is trained in knots, the same indication in mph may result in undershooting, a stall about 15 ft in the air, or at best a quick blast of power.

This is a double-edged-sword type of problem, because it is possible to become trapped into approaching too fast or too slow, depending upon recent training or type of equipment flown. And the blame for errors arising from misreading the indicated airspeed cannot be placed entirely upon the pilot. When the going gets a bit tough or fatigue takes over, the natural tendency is to revert to familiar values; so if trained in mph readout, a pilot will read all indicated values as mph—but when the instrument is calibrated in knots, accidents may occur—and vice versa.

As a result, airspeed indicators can be the source of considerable confusion in establishing proper approach and climb speeds. If thinking must be switched from mph to knots, the dial should clearly carry the word "knots" in solid letters plus the simple formula "mph = knots × 1.15." Hopefully, any qualified pilot can then add 15 percent to get speeds into familiar mph terms. Checking a pilot in handling a new model is required procedure and of obvious benefit, while providing some peace of mind for the FBO (fixed base operator) as well, but if the rental pilot may not fly that same airplane with its "knots" airspeed indicator for a month or more during which time he or she flies "mph" airspeeds, having the instrument clearly show that it is reading in knots would be very helpful. Of course, the reverse is also true; all mph airspeed instruments should clearly carry the abbreviation "mph" plus state that "mph = knots × 1.15."

WIND SHEAR

There is another cause of under- and overshooting which we should review—the newly recognized and very important local wind shear condition. We have all been trained to approach around 1.3 to 1.4 times stalling speed, or slightly higher if variable gust conditions lie ahead. Unfortunately, ground wind information is not normally available at uncontrolled airports, and may not be given by unicom operators unless an unusually severe situation exists at the airport. But even major airline terminals equipped with all available meteorological equipment occasionally experience wind shear intensities capable of destroying large aircraft operated by highly skilled crews.

If this is the case, what chance does a small airplane have in similar conditions? The answer is very little—so the situation is one to be recognized and avoided if possible as the following illustrations indicate. These sketches, based upon an excellent article presented in Shell Aviation News (Ref. 9.3) and a re-

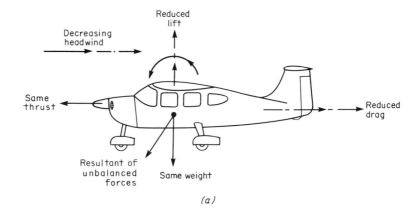

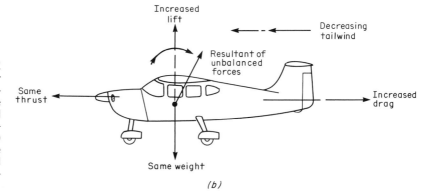

FIGURE 9-2
Effect of horizontal wind shear upon the flight path. (a) Decreasing headwind. Airplane shows a reduction in indicated airspeed and goes down. (Similar to increasing tailwind.) (b) Decreasing tailwind. Airplane shows an increase in airspeed and climbs. (Similar to increasing headwind.)

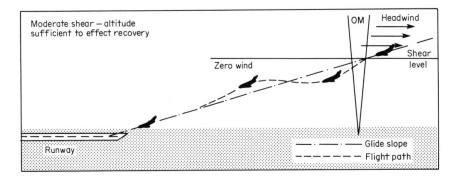

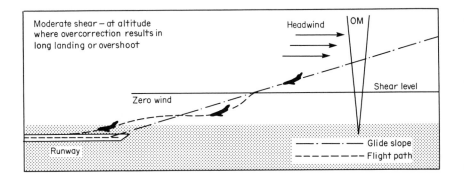

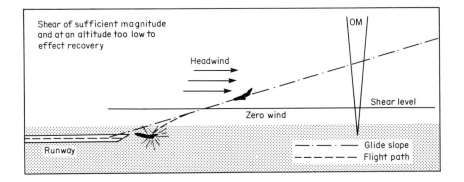

cent FAA circular (Ref. 9.4), have been included to show what wind shear is and how to avoid the dangers of severe wind shear conditions.

Basic to all this is an understanding of what happens to an airplane experiencing wind shear of decreasing headwind or decreasing tailwind type. The resultant unbalance of forces shown in Figure 9-2(a) and (b) modify the flight path toward the direction of the force vector. In the case of decreasing headwind, the airplane will change attitude from an apparent decrease in airspeed and go down; with decreasing tailwind the effect is similar to flying into an increasing headwind, so the apparent airspeed over the wings will increase

FIGURE 9-3 (a) Recovery from headwind shear:
· Loss of indicated airspeed is equivalent to shear value.
· Lift is lost, airplane pitches down, drops below approach glide path.
· Pilot applies power to regain airspeed, pulls the nose up, and climbs back onto approach path.
· Pilot probably overshoots the approach glide path and target airspeed but recovers and lands without difficulty.

FIGURE 9-3 (b) Overshoot from headwind shear:
· Loss of indicated airspeed is equivalent to shear value.
· Lift is lost, airplane pitches down, drops below approach glide path.
· Pilot applies power to regain airspeed, pulls the nose up to climb back onto the approach path. Nose-up trim may have been used.
· When airspeed is regained, thrust required is less than needed with previously existing headwind.
· Thrust is not reduced as quickly as required, nose-up trim compounds problems, airplane climbs back above approach path.
· Airplane lands long and fast.

FIGURE 9-3(c) Undershoot from headwind shear:
· Loss of airspeed is equivalent to shear value.
· Lift is lost, airplane pitches down, drops below approach glide path.
· Pilot applies power to regain airspeed, pulls the nose up to climb back onto the approach path.
· Airplane is in high-drag approach configuration at low altitude. Increased AOA produces only a momentary increase in lift accompanied by a large increase in drag as the maximum value of the lift/drag ratio is exceeded. The result is a momentary arrest of descent with decreasing airspeed followed by a large increase in the already high descent rate.
· Pilot's only chance to save the landing is to pull back on the stick and apply full power.
· If this response is too late, the airplane will crash short of the runway.

and the airplane climbs—unless the angle of attack can be instantly decreased in proportion to increased airspeed.

As you probably know only too well, when on final with the runway end boresighted for a good landing, any variation in apparent airspeed over the wings will change the glidepath. Any momentary increase in indicated airspeed will reduce the descent rate, while any sudden reduction in indicated airspeed increases descent. Although Figure 9-3 shows the effect of decreasing headwind upon an airplane in instrument conditions, the results are the same for any type of approach.

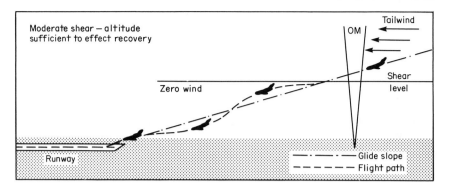

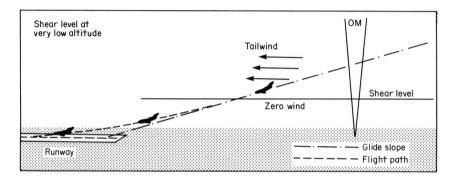

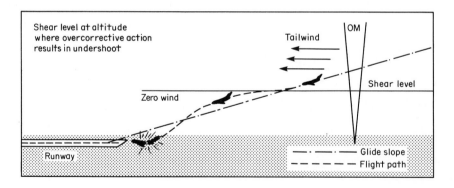

Note that when decreasing headwind shear is of low value or is encountered high enough above the surface, an experienced pilot can effect recovery without too much difficulty. Fortunately, recovery in attitude, speed, and altitude is somewhat simpler for our small aircraft than for a large commercial transport, since our wing loadings and airplane mass are so much lower. But when recovery cannot be accomplished in the distance or altitude available, an overshoot or undershoot develops as shown in Figure 9-3(b) and (c).

As would be expected, decreasing tailwind produces the opposite effects as shown by Figure 9-4, but again the final result of an improper or impossible

FIGURE 9-4(*a*) Recovery from tailwind shear:
· Increase in airspeed is equivalent to shear value.
· Lift increases, airplane pitches up and rises above the approach path.
· Pilot reduces power to reduce airspeed, noses over to return to approach path.
· Pilot now needs more thrust than when flying in the tailwind condition.
· Pilot probably overcorrects and airplane descends below approach path with speed increasing beyond target value, but recovery can be made and the landing is successful.

FIGURE 9-4(*b*) Overshoot from tailwind shear:
· Increase in airspeed is equivalent to shear value.
· Lift increases, airplane pitches up and rises above the approach path.
· Increased speed and height above the normal flight path result in a fast and long landing.

FIGURE 9-4(*c*) Undershoot from tailwind shear:
· Increase in airspeed is equivalent to shear value.
· Lift increases, airplane pitches up and rises above the approach path.
· Pilot reduces power to reduce airspeed to target value, noses over to return to approach path. Nose-down trim is likely applied.
· Airplane decelerates in response to reduced thrust, and target airspeed is regained with need for more power than was required in the earlier tailwind portion of the approach.
· Airplane is in a high-drag configuration with insufficient thrust. Descent below approach glide path occurs at a critically low altitude. Nose is pulled up sharply, increased AOA produces only a slight or momentary increase in lift accompanied by a large increase in drag as maximum lift/drag ratio values are exceeded. The result is a momentary arrest of descent with rapidly decreasing airspeed, followed by a large increase in an already high descent rate.
· Pilot's only chance to save the landing is to pull back on the stick and apply full power.
· If this response is too late, the airplane will crash short of the runway.

recovery is an undershoot or overshoot landing. As indicated by the caption of Figure 9-2(*a*) and (*b*), an increase in tailwind velocity will cause a reaction similar to decreasing headwinds, while increasing headwinds create the same situation experienced with decreasing tailwinds. Naturally, there are variations from these four basic encounters, with possible changes in velocity gradients including descent through layers differing in both velocity and wind direction. As any instrument pilot can testify, these conditions can make for a most difficult approach.

For most horizontal shear values, flying an AOA indicator approach will

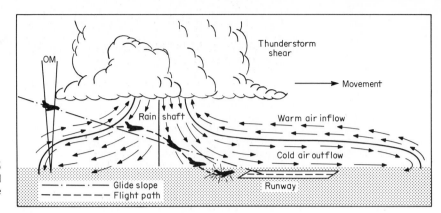

FIGURE 9-5
Horizontal and vertical wind shear in the vicinity of an active thunderstorm.

compensate for variations in local wind strength; so once again this instrument can provide increased flight safety by serving as a useful basic reference. But neither an AOA nor most small aircraft can correct for vertical shear gusts of the intensity found near squall line or thunderstorm activity.

Figure 9-5 presents a particularly clear and graphic explanation of why thunderstorms are as dangerous around an airport as they are along an airway. For the situation shown, the combination of tailwind plus downdraft have turned a potential overshoot into an undershoot. But note that if the thunderstorm had been approaching the airport from the right, an airplane on final would have encountered cold headwinds or rising warm air currents that might result in overshooting the runway. So beating a thunderstorm to the airport could become a shallow victory; turning 180° to seek another airport would be the preferred and safer decision under such conditions.

Because it is not possible to modify local weather to suit our pleasure, a certain number of approach and landing accidents will always remain with us. But we can do something about undershoot and overshoot accidents that are not weather-related. These can be reduced through:

1. Improved pilot training in stall recognition and final approach technique
2. Stall-preventing basic aircraft design features outlined in Chapter 3
3. Use of the AOA indicator

Now that we are down on the runway, for better or for worse, let us next see what effect improved flight technique might have in reducing so-called pilot error accidents.

REFERENCES   9.1   (a)   *Annual Review of Aircraft Accident Data—U. S. General Aviation Calendar Year 1970*. National Transportation Safety Board Report NTSB-ARG-74-1, 1974, p. 7, table D.
(b)   *Annual Review . . . 1971*. Report NTSB-ARG-74-2, 1974, p. 9, table 4.

9.2   *Flightcrew Coordination Procedures In Air Carrier Instrument Landing System Approach Accidents.* National Transportation Safety Board Report NTSB-AAS-76-5, August 1976.

9.3   Capt. J. T. Fredrickson, Wind Shear—An Update, *Shell Aviation News,* no. 439, 1977.

9.4   FAA Advisory Circular, AC No. 00–50: *Low Level Wind Shear,* April 8, 1976.

10 TOWARD SAFER FLIGHT TECHNIQUE

Throughout this book it has been necessary to relate flight procedures and handling qualities to various design features intended to reduce pilot error accidents. As a result, many phases of flying have been reviewed in earlier chapters; however, to assist in coordinating and improving pilot technique as well as to provide a future reference source, it seems advisable to summarize prior operational discussions in this one chapter.

KEEPING SCORE

Interestingly enough, accident data reviewed by regulatory agencies in the United States agree closely with similar European studies conducted in England. In both cases weather was found to be the largest single factor in 40 percent of all accidents, closely followed by stall/spin causes. And as noted in Figure 8-1, the accident rate increases the further flight progresses along the flight profile.

The majority of weather accidents occur during the in-flight or cruising phase, generally in areas with low clouds and reduced or no visibility. Probably the main cause for the continually increasing number of accidents of this type lies in the growing number of small aircraft with navigation equipment encouraging pilots to "continue just a bit further to see if conditions will improve." Unfortunately, these are frequently final words if onboard equipment exceeds pilot competence.

Recent NTSB data indicate the largest percentages of all accident causes are the result of pilot error, regardless of pilot rating or experience; and the values remain fairly constant year after year. For example, 82 percent of general aviation accidents are pilot-related, while 70 percent of National Business Aircraft Association (NBAA) corporate accidents have been traced to pilot error, as have been 51 percent of all air carrier (airline) accidents. Further, there is a continuing incidence of high-time pilot accidents with only a few hours in type; typically an 11,000-hr pilot with only 20 hr or so total time in the airplane model involved.

Please note that these weather- and pilot error–related statistics are not contradictory to Figures 1-1 and 1-2. The pilot error and weather percentages reported in those summaries cover only selected accidents over which we can exercise some degree of design or systems improvement, and so do not include all accidents of either type for the reporting period. However, the total accident and cause/factor values listed in these figures do include all accidents reported for the year given.

Who owns these airplanes and how are they equipped? Based upon a recent FAA report, the 1976 general aviation fleet was owned as follows: 41.8 percent by professional, technical, and similar workers; 30.5 percent by managers and administrators; and 18.7 percent by craftsmen, sales, and clerical pilots. The remaining 9 percent was spread over various professions and interests.

Most of these were single-engine aircraft with 47.2 percent of the total general aviation fleet having one engine plus four or more seats, while 37.3 percent were single-engine and contained three seats or less. The 15.5 percent balance would be a mixture of twins and helicopters. These general aviation airplanes were well equipped with 77.7 percent having radios and 76.8 percent including VOR (very high frequency omnidirectional range receiver). And this equipment is one major reason for pilots pushing ahead in marginal weather, until they eventually become another unwilling member of the growing enroute accident group.

ACCIDENT PREVENTION

A thorough weather briefing before starting off cross-country, combined with some caution plus a realistic evaluation of personal flying skill, would prevent a great number of unnecessary weather accidents. Rather than entering a long discussion of flight procedures, which are amply covered by the aviation press, the following suggestions are offered as items for serious consideration by active pilots. We could considerably improve flight safety and reduce the general aviation accident rate if every pilot would do as follows:

1. Thoroughly prepare and document all routes for every cross-country flight. Reference 10.1 provides comprehensive considerations and suggestions for proper operation.

2. Take 10 min before every departure to make a thorough preflight inspection of the airplane, even if flown previously the same day, and look into all tanks (fuel, oil, hydraulic) to personally check the actual fluid levels.

3. Turn the propeller over by hand 3 or 4 times during preflight if the airplane has been standing for half a day or longer, for reasons discussed previously in Chapter 8.

4. Maintain accurate and current weight and balance calculations and use them to keep all operations within approved c.g. limits and design gross

weight. Reference 10.2 contains basic information for the preparation and use of aircraft loading charts.

Also, be sure you know what is placed aboard any airplane you will be flying. A Cessna 207 was recently destroyed and the pilot killed because the airplane was loaded to gross weight by one cargo crew, and then the balance of a similar cargo was subsequently placed aboard by a second crew. As a result, the pilot attempted takeoff at 790 lb over gross and with the c.g. behind aft limit. Naturally, this fatal accident was blamed upon "inadequate preflight preparation," with failure to check cargo weight and location being the real problem. Interestingly enough, if this load had been aboard an airplane that could not have been loaded aft of rear c.g. limits, the pilot might have retained sufficient control to abort the takeoff. The general configuration shown by Figure 6-18 offers this safety.

5. Prepare a complete checklist *and use it* for the different phases of flight—particularly for all pre-takeoff and landing procedures. This practice is faithfully followed by airline and military pilots, and for good safety reasons. The checklist should be mounted in a convenient spot on the instrument panel; checklists are frequently ignored because they are not readily available or are hidden behind control wheels, engine controls, etc. The use of a comprehensive checklist is most important for retractable gear aircraft and critical for amphibian operation, where opportunity exists to land on land gear up and on water with the gear down. (See Chapter 6, page 99.)

6. Install and use a shoulder harness for each occupant. (See Chapter 6, pages 86–89.)

7. Carefully check runway length and surrounding terrain for density altitude and takeoff weight conditions. To check temperature and altitude effect upon takeoff distance and ROC, use a Denalt computer (available from the Superintendent of Documents) for either fixed-pitch or constant-speed propellers. The effect of temperature on ROC is shown by Figure 10.1.

Density altitude conditions may be obtained from Figure 10-2 once temperature is known and the altimeter is set to barometric reading 29.92 to find pressure altitude. The density altitude based upon ambient conditions is found by going vertically up from outside air temperature until reaching the pressure altitude line, and then moving horizontally left to read approximate density altitude. For example, at 90°F ambient temperature and 4000-ft pressure altitude, the density altitude is 6900 ft. Your Denalt computer shows how much this will *increase* the takeoff run and *decrease* the ROC. Use it to avoid becoming another takeoff statistic due to "improper preflight preparation."

Variations in gross weight also affect takeoff and climb as shown by Figure 10-3, based upon typical airplane flight manual curves. Be familiar with those for your own airplane and keep them in the cabin for ready reference.

8. Avoid zero-zero takeoffs when weather is right down to the runway or nearly so. A little thought will reveal why a zero-zero takeoff can be most

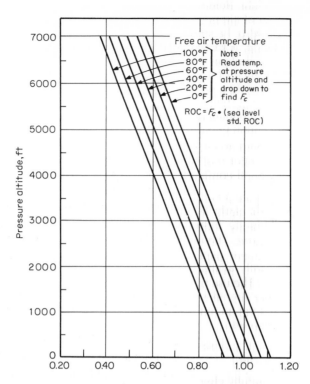

FIGURE 10-1
ROC correction for altitude and temperature.

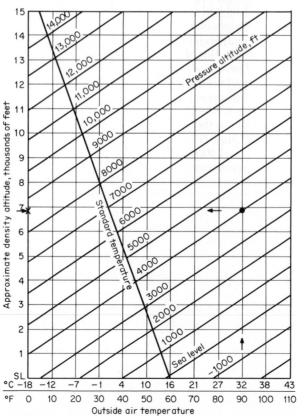

FIGURE 10-2
Density altitude variation with temperature and pressure altitude (29.92 altimeter setting). For the example shown, at 90°F and 4,000-ft pressure altitude the density altitude will be approximately 6,900 ft.

156

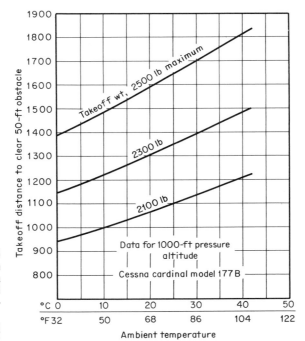

FIGURE 10-3
Takeoff distance variation with weight and temperature. Data for Cessna Cardinal model 177B at 1000-ft pressure altitude.
Cessna Pilot's Operating Handbook as replotted

hazardous; if the engine stops, where should you go? For this simple reason, have at least 300-ft clear altitude before attempting such an IFR takeoff, and then go only if you must. If the engine fails below 300 ft, the ground will be visible; and if failure occurs in the clouds shortly after takeoff, a return glide to the field and some decision concerning landing area may be possible after breaking out at 300 ft. The statistics are quite simple: powerplant failure does occasionally occur during takeoff. Reduce the adverse odds in the unlikely event an emergency landing becomes necessary by never taking off unless there is sufficient ceiling to safely perform a shallow bank toward a landing site.

9. Use an AOA indicator for proper climb-out and approach speed as well as for optimum cruise trim. This instrument is particularly useful for short-field operation as previously discussed.

10. Obtain crosswind takeoff and landing instruction in different wind intensity, angle, and gust conditions under direction of a qualified instructor; then get out and practice crosswind work. (See Ref. 10.3 for detailed suggestions and recommended procedures.)

11. Obtain ground school plus actual flight instruction to the extent possible concerning the effects of heavy rain, snow, slush, and ice upon runway surfaces. This training does not lend itself to simulator practice; but for safety, a qualified instructor could demonstrate procedures at fairly low speed along remote taxiways or ramp areas free of other aircraft or obstructions.

12. Receive thorough instruction when getting a water rating, and realize that separate instruction is required for floatplane and hull-type boat operation; the handling procedures are different for each type of seaplane. Also, be sure to receive an equally thorough water check when operating any new seaplane; the various models behave differently from one another, so some dual instruction is recommended.

   If you switch from wheels to floats, don't duplicate the mistake of one pilot who had just installed floats for summer operation. When the airplane developed a balky engine over Adirondack lake country, he proceeded to land in a plowed field, completely forgetting the ship was now on floats instead of wheels, with water landing fields in all directions to compound the error.

13. Receive skiplane instruction before operating from snow. A snowplane rating should be required, similar to a water rating, including checkout in different snow and ambient temperature conditions. Be careful when operating from snow- or ice-ridged runways; with such surface conditions a skiplane can be thrown into the air before reaching flight speed. The return to earth can be damaging to an airplane and also to a pilot's back, so inspect snow- or ice-covered runways prior to takeoff. To the extent possible, the same caution applies to skiplane landings.

14. Be familiar with the use of downward vision angle cabin reference to determine the distance from frontal action, thereby signaling approaching bad weather in time for a safe 180° turn back to VFR conditions. As shown by Figure 10-4, the downward vision angle may be established quite easily. From 3000 ft an 11° angle will permit sighting the ground 3 mi away. If the surface disappears in haze, fog, or rain, turn around; from 3000 ft above ground level (AGL) you will have 3 mi in which to complete a safe 180° turn. From 4000 ft, about 4 mi, etc. Other downward angles and flight altitudes provide warnings in similar proportion. For example, from 3000 ft AGL a 5.5° downward angle would contact the surface 6 mi ahead of the airplane, providing additional safety and time to turn back to clearer weather. Check your airplane and mark the windshield at the angle providing useful downward sighting; note that high-performance aircraft would be better served by a shallow sighting angle, which provides an earlier warning for safe evasive action.

15. Receive ground and flight instruction to ensure familiarity with wake tur-

FIGURE **10-4**
Determining the downward vision angle θ.

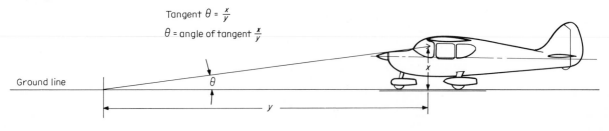

Tangent $\theta = \frac{x}{y}$

θ = angle of tangent $\frac{x}{y}$

Ground line

bulence, downdraft, and wind shear conditions. (See detail descriptions in Chapters 9 and 12.)

16. Obtain thorough instruction in powerplant operation, stall recognition, spin entry and recovery procedures, flying in turbulence (ground school and flight work), radio procedures, local and standard pattern procedures, and radio communications—VFR and IFR as applicable. Spin instruction was required for all military pilots during World War II and has been credited with saving many lives.

17. Be familiar with the accuracy and limitations of all onboard navigation, communications, and flight instrument equipment used for IFR operation. Also, be sure that when dual units are installed they are both functioning correctly when departing. The DG rate of precession is particularly important for procedural turns and ADF approaches and should not exceed 10° for 30 min.

18. Never intentionally plan to fly in weather exceeding current levels of pilot proficiency or equipment capability (VFR or IFR). (See Ref. 10.4 for background.)

    One form of weather that can be seen and avoided, at least in daylight, is a developing or active thunderstorm. The wind shear forces caused by gust action within a thunderstorm can tear a military airplane apart, so our light aircraft should never be exposed to such weather. Although no one willingly enters an active thunderstorm, trouble occurs when flying IFR directly into unreported imbedded thunderstorm cells, running VFR into unpredicted line squall activity, or approaching too close to visible thunderheads.

    While not so well known, snow clouds and showers can also contain considerable turbulence, frequently rivaling thunderstorm violence. Anyone regularly flying the Alleghenys will be aware of these conditions, because it is impossible to fly seriously in that area without contacting storm activity of all sorts at all times of the year. When heavy weather is predicted, stay clear of the area by skirting around the front or by remaining on the ground.

    If caught in unpredicted turbulence of some severity, immediately reducing airspeed to just under twice stalling speed at flight weight will keep your airplane together provided you do not overcontrol to maintain attitude and altitude.

    This inherent safety feature is at our command because airframe loading is directly related to the squared ratio of flight speed divided by stalling speed. This means that an airplane flying at 120 mph and stalling at 60 mph will not develop a structural loading greater than $(120/60)^2 = (2)^2 = 4$ g. And all FAA approved small aircraft are designed and tested to 3.8 g with a 1.50 factor of safety = 5.7 g ultimate loading. If you wish a somewhat greater safety margin, throttle back to 110 instead of 120 mph. However, flying too slowly can result in getting undesirably close to a full stall if and when horizontal gusts are encountered.

    For example, if the above flying speed is reduced to 90 mph, a 35-mph gust striking from behind would suddenly reduce airspeed over the wings to 55 mph (90 − 35 = 55). Since this is 5 mph below our noted stalling speed of 60 mph, the airplane can no longer fly and down we go.

So if caught in unexpected turbulence with no ready exit possible: reduce cruising speed to slightly below twice the stalling speed of your airplane, do not fight the controls, and do not panic. If you follow these simple emergency rules, any well-maintained airplane will stay together and be able to take more than you can, or care to.

19. Use oxygen when flying above 10,000-ft pressure altitude for periods greater than a half hour. This is especially critical after dusk when oxygen deficiency can reduce night vision.

20. Leave retractable landing gear extended during takeoff until so far down the runway it would be impossible to land straight ahead on the field if the powerplant should fail. Pulling the gear up too early is an invitation to a takeoff accident caused by gusty air or a momentary decrease in engine power. Use the checklist to be sure gear is down when landing on land (and gear up for an amphibian landing on water).

21. Report all damage incurred during aircraft rental or lease. The life of the next pilot to fly may depend upon this level of honesty and integrity. This procedure should be an FAA enforced regulation.

22. Never mix alcohol or business and personal problems with flying. They may have their place, but it is not with you in an airplane.

23. Remain alert at all times for other aircraft, particularly around airports, and for bird strikes during migration periods. Having a passenger casually note a flock of geese passing below can be a disturbing experience. Bird strikes are more common than generally realized, with the maximum number of accidents regularly occurring during October (Ref. 10.5).

24. Consider the importance of survival gear when flying in northern climates during the late fall and winter months. Some worthwhile suggestions and procedures have been collected in Ref. 10.6, including those for water survival.

25. Possibly most important of all recommendations: remember to remain calm during an emergency if one develops. The odds are with you if you don't panic. Some useful emergency procedures may be found in Ref. 10.7 along with some quieting suggestions.

In view of the number of commercial pilots and their percentage of accidents indicated by Figure 1-4, consideration should be given to requiring an instrument rating as a prerequisite for a commercial pilot's license. As more pilots fly for hire, adverse weather exposure increases along with the temptation to complete business flight schedules and other tasks regardless of changing weather conditions. In many cases, proper instrument training would have made the difference between a safe flight and another commercial pilot accident.

While on the subject of IFR operation, one helpful procedure I have used is that of clearly writing down my flight plan as filed. This data may then be followed readily as the clearance is received and easily corrected where revised by air traffic control (ATC). This simple step assists in getting the clearance correctly set down as well as permitting fast readback to control.

As a final thought, while there is little we can do to prevent midair heart attacks or other sudden sickness, passengers who frequently fly with you (wives or husbands and friends) should be flight and radio checked through the local pattern and landing procedures—and practice at least one landing a month. The various pinch-hitter and similar familiarization courses provide excellent instruction for handling an airplane during such emergencies. Waiting until trouble occurs is a poor way to dodge the possibility since a little preparation can save many lives and deserves serious consideration for everyone's safety.

REFERENCES

10.1 FAA Advisory Circular, AC No. 61-84: *Role of Preflight Preparation,* April 11, 1977. Available from the FAA, Washington, D.C. 20591.

10.2 Don Hewes, Construction of a Simple Loading Chart for Your Airplane, *Sport Aviation,* July 1978, p. 32.

10.3 David B. Thurston, *Design for Flying.* McGraw-Hill Book Company, New York, 1978, chap. 5.

10.4 FAA Advisory Circular No. AC-00-6A: *Aviation Weather,* March 3, 1975. Available from the Superintendent of Documents, Government Printing Office, Washington, D.C. 20402. Order Stock No. 050-007-00283-1. Price $4.55 in 1978.

10.5 FAA Advisory Circular, AC No. 20-49: *Analysis of Bird Strike Reports* . . . , July 27, 1966. Originally issued by the FAA, Washington, D.C. 20591.

10.6 (a) Windchill, *FAA Aviation News,* February 1969, p. 6. (Includes survival vs. exposure time charts.)

(b) *Tips on Winter Flying.* Available from the FAA, Washington, D.C. 20591.

(c) Larry Collier, Objective Survival, parts 1, 2, and 3 (mountain, desert, and water survival), *Aero Magazine,* January, February, and March 1976.

(d) Joseph R. and Antonia Novello, M.D., The Will to Survive, *Aero Magazine,* November 1976, p. 60.

(e) Glen Tabor, The Killing Assumption, *The AOPA Pilot,* October 1977, p. 84.

(f) J. D. Greiner, SG ONBD, *The AOPA Pilot,* June 1978, p. 46.

(g) J. R. Williams, Ditching to Survive, *The AOPA Pilot,* May 1975, p. 35.

10.7 Robert N. Buck, Emergencies, *Business and Commercial Aviation,* November 1975, p. 48.

11
SAFETY THROUGH MAINTENANCE

While any discussion concerning aircraft maintenance may seem out of place in an aircraft design book, these subjects are related because a well-designed airplane includes ready access for inspection of equipment and critical structural areas. If suitable inspection openings are not provided where required throughout the airframe, annual inspection costs will be higher than necessary; more importantly, components in need of inspection or repair may never be noted at all until something fails and another accident is recorded. Provision for service and inspection access throughout the airframe is an important safety feature of aircraft design; ease of maintenance really starts on the drafting board as the preliminary design configuration becomes finalized.

AIRWORTHINESS DIRECTIVES

Rapid incorporation of FAA Airworthiness Directives (ADs; see page 21) is an essential part of continuing airframe maintenance. Occasionally, the aircraft or component manufacturer pays for part or all of the AD costs, but usually the owner foots the bill. While putting off AD work may seem to save money, it may instead cost your life if some critical inspection or repair has been neglected.

The fact that similar model aircraft have been flying without benefit of the required AD repair is not a valid reason or argument for ignoring AD work. It is quite likely that some critical structural member has fatigued or just plain failed under service use, and the AD has been issued to prevent any similar airplane from suffering the same accident. Since ADs are backed by engineering analysis and judgment tempered by prior operational experience, they should be acted upon as required when issued.

It is now possible for an owner to perform part of the annual inspection on a controlled, continuing basis under supervision of a local airframe and power-plant (A&P) mechanic, an enlightening and frugal procedure, with helpful and general background for much of this work clearly provided by the FAA in Ref. 11.1.

RECORDS Good maintenance and modification records are most important, not only to prove certain work has been completed but also for future reference and to satisfy FAA regulations. Since every aircraft owner is responsible by law for making proper entries in the aircraft and engine log books, the recording procedures outlined by the FAA (Ref. 11.2) should be of great assistance—they even include a suggested format for keeping AD compliance records. Legible and up-to-date service records are as necessary in many ways as proper aircraft maintenance; both are essential for flight safety, FAA licensing, and minimum possible insurance premiums.

While on the subject of records, aircraft weight and balance entries should be given the same priority as is required for maintenance and modification work. Although this subject is thoroughly covered by Refs. 11.3 and 11.4 and will not be reviewed in detail here, from the safety standpoint every pilot should realize that all changes to an airplane which alter the originally purchased configuration or equipment will also affect the weight and probably move the c.g. as well. Such modifications may be minor or major, but either way airplane log entries should correctly reflect the current airplane weight and c.g. position prior to loading people, fuel, oil, and baggage. And this information should be readily available so that no one flying the airplane will have to go through a discouraging and time-consuming hassle trying to find or guess the latest weight and balance values.

I have rented aircraft with no weight records aboard, with weight entries made as individual item changes which were not summarized into current configuration weight and balance data, with weight modification information randomly scattered through the log and not arranged in chronological order, and with latest weight data bearing little relation to equipment on board. When a pilot is faced with a critical gross weight or density altitude takeoff, a marginal or erroneous weight and balance statement could well result in another "pilot error" accident due to "failure to obtain flying speed" before the crash occurred.

Weight and balance records should be clearly dated in the aircraft log, and, ideally, should also list all removable equipment such as avionics and instruments included in the current weight statement. Through this procedure anyone can immediately determine whether the weight and balance data are properly up to date and also find the actual flight weight. Note that all this record-keeping is not entirely a basic safety item; accurate weight and balance data is also a legal requirement of FAA flight regulations and should be available *in the airplane* at all times.

As a final thought about weight in general, remember that from the time a well-designed airplane first flies, everything added to it will contribute to increased empty weight, reduced performance, and greater cost. This also holds true throughout the life of the airplane; and since useful load as well as takeoff and climb performance decrease with increased weight, both utility and safety suffer from every addition of unnecessary fixed weight. So be sure anything permanently installed is really required; unnecessary items cause increased maintenance while reducing performance.

INSTRUMENT CHECKS
Anyone seriously flying IFR, and there is really no other way to fly by instruments, should maintain all avionics, navigating equipment, autopilot if installed, and instruments in perfect working order. This means dual communication and navigation systems must both be working when there are two of each—not one set functioning while the other is out for repair. After all, the remaining set could fail in flight, which is why two systems were originally installed.

DG drift should be an absolute minimum, and the instrument changed or repaired if drift exceeds 10° in 30 min. Quite frequently there is no time to worry about gyro drift correction or no opportunity to reset properly if a solo IFR approach becomes necessary in turbulent air. And a badly drifting DG is useless on an ADF approach or as backup for a long localizer leg inbound from an approach fix beacon.

The same applies to autopilots. If they work correctly, fine. If they do not, they may be very dangerous in rough going and do more harm than good. Autopilots should be frequently checked for operation in VFR weather, so they will be capable of providing support when needed.

Of course, maintaining radios, gyros and other flight instruments, and the autopilot in proper working condition also represents more than a safety item; IFR flight is not legal unless all required equipment is correctly adjusted and operating within specified limits. Since this same equipment can also be helpful and time-saving during VFR operation, there is no practical (or legal) reason not to keep it that way at all times along with supporting records.

PREFLIGHT CHECKS
Although really a flight item, proper preflight check of the airplane and equipment is also a maintenance and safety procedure. I have occasionally found fairing screws backed out to interfere with elevator movement—twanging away on the metal covering as the elevator moves up and down, a blown exhaust gasket, cracked exhaust stacks, cut tires, and a loose propeller spinner plus many minor items.

A thorough preflight inspection may detect loose aileron or flap hinge and pushrod attachment bolts, cracked support brackets or skins, damaged tip lights preventing proper identification (safety) in haze or at night, etc. These are all items that require maintenance support for safe operation prior to flight. As a matter of personal safety, do not neglect a thorough preflight inspection, including a check of actual oil and fuel levels as well as their freedom from water or other foreign material.

Fuel contamination raises its ugly head more often than it should, but less frequently than it used to before fuel delivery was so well screened and strained as it is today. The principal contaminant is water arising from condensation formed in partially filled aircraft fuel or storage tanks, and from outside moisture such as rain or snow entering around filler caps or through vent systems. Water entry of this type can occur during seaplane operation, requiring great care in properly locating the fuel tank filler and vent outlets as determined by flight test.

We are also developing an even more dangerous though less frequent form of fuel contamination today by filling aircraft tanks with the wrong fuel. A number of serious accidents have been reported here and in Europe as the result of filling gasoline tanks with jet fuel. So check your tanks when filling up at a multiple pump or field system. This occurrence could be prevented by using specially shaped fuel nozzles and matching tank filler openings for each different type of fuel (see Chapter 5, page 70).

OIL ANALYSIS   Many owners have their engine oil regularly checked by spectrometric oil analysis to determine whether bearing or other metal particles have entered the oil system through engine wear. This is a useful form of preventive maintenance contributing to flight safety by indicating potentially serious engine wear before some mechanical or structural failure occurs in flight. Such analysis is possible through detection of extremely small particles formed at the first stage of engine wear. By determining the type of metal in the oil sample, it is frequently possible to identify the particle source and so locate the exact area of engine wear, which can save tearing down the entire engine as noted in Ref. 11.5, and may also prevent an engine failure in flight. As a result, regular oil analysis is a useful and highly recommended maintenance safety procedure.

SUMMARY   By way of closing with the same theme discussed at the beginning of this chapter, it is most important that maintenance and service access be thoroughly considered as a new design progresses. Not long ago a leading aviation magazine carried an indictment of a new model business jet that carried spare fuses in the tailcone, of all places. If a spare fuse was urgently required during flight, replacement would require some long reach back from the flight deck at 600 mph.

With some thought, I'm sure anyone could better that level of accessibility. Of course, the easier it is to perform a routine inspection, the more frequently it will be made and the earlier required maintenance can be detected, while necessary work can be completed at lower cost. Adequate maintenance accessibility is an important design consideration directly contributing to increased flight safety and lower operating cost throughout the life of an airplane.

REFERENCES   11.1   (a)   FAA Advisory Circular, AC No. 43.13-1A: *Acceptable Methods, Techniques and Practices—Aircraft Inspection and Repair;* April 17, 1972. Available from the Superintendent of Documents, U.S. Government Printing Office, Washington, D.C. 20402. Order Stock No. 050-011-00058-5. Price $3.70 in 1978.

          (b)   FAA Advisory Circular, AC No. 43.13-2A: *Acceptable Methods, Techniques and Practices—Aircraft Alteration;* June 9, 1977. Available from U.S. Government Printing Office. Order Stock No. 050-007-00407-9. Price $2.75 in 1978.

11.2   FAA Advisory Circular, AC No. 43-9A: *Maintenance Records; General Aviation Aircraft;* September 9, 1977. Available from the FAA, Washington, D.C. 20591.

11.3   David B. Thurston, *Design for Flying.* McGraw-Hill Book Company, New York, 1978, chap. 9.

11.4   FAA Advisory Circular, AC No. 91-23A: *Pilot's Weight and Balance Handbook;* rev. 1977. Available from Superintendent of Documents, U.S. Government Printing Office, Washington, D.C. 20402. Order Stock No. 050-007-00405-2. Price $2.30 in 1978.

11.5   Malte Lukas, Oil Analysis a Proven Aviation Maintenance Tool, *Aviation,* September 1977, p. 58.

12

AIRWAYS AND AIRPORTS

THE PROBLEMS   While not related to aircraft design improvements or possible deficiencies in the sense or degree we have been considering up to this point, airway and airport problems as listed in Figure 1-1 regularly average about 4.50 percent of all cause-and-factor accidents. With airways facility incidents averaging some 30 percent of this total, it becomes apparent that some refinement of our procedures is just as warranted as a closer review of aircraft design features. Fortunately, these accidents are rarely fatal, although they do not improve our private flying safety record or public image.

If we also add in the number of accidents due to high obstructions located around airports, the total jumps to a fairly steady annual average of 1050 accidents, with about 15 percent being fatal. These are unacceptable rates, indicating need to either clean up the pattern and approach areas for about a 5-mi radius around active airports or relocate the airports. Although neither of these solutions seems particularly practical or acceptable for existing airport facilities, they should be given consideration when selecting a site for new airports, whether small or large.

Following high-obstruction collisions the next ranking cause involves airways facilities. This includes accidents arising from improper IFR procedures, such as difficulty with or misunderstanding communications, etc. Many IFR procedures could and should be simplified to accommodate the rated pilot who maintains a minimum acceptable level of instrument flight proficiency but does not make daily IFR approaches. This category covers a major percentage of those private pilots who are IFR rated.

When flying on relatively remote airways or away from terminal control areas (TCAs) and busy airports, simpler holding patterns or basic loitering procedures could be used. For the nonprofessional pilot IFR procedure should be a safety operation, not a test of skill and endurance. Although I fully realize this is a controversial and touchy subject, please remember we are discussing the pri-

169

vate pilot who holds a current instrument rating primarily as a flight safety and proficiency procedure and not to provide income or competetive advantage.

When aircraft are approaching an omni transmitter or entering the airport traffic pattern, opportunities for midair collisions tend to increase. Fortunately, these are few in number due to pilot vigilance in congested areas plus the fact that most of the country is open space containing few aircraft. But the results are heavily fatal when such accidents occur. Visibility is an important factor in preventing collisions of this type, since the best swivel-neck is of little benefit if a wing or cabin top obstructs the view up, down, sideways, or, equally important, aft.

As personal aircraft traffic increases at airports scheduling DC-9 and larger equipment, the problem of wake turbulence becomes increasingly important and dangerous to small airplanes. Landing too soon after large aircraft, particularly when on or near their approach glidepath, can provide some exciting and frequently fatal low-level aerobatics.

The huge vortices flowing off the wings of large aircraft flying at low airspeeds (high-lift condition) contain sufficient energy to overturn small single- and twin-engine airplanes.

The same is true to a lesser degree if the takeoff climb path closely follows that of a large airplane. During climb, a transport is normally operating at an airspeed slightly higher than during the approach but still in a high-lift configu-

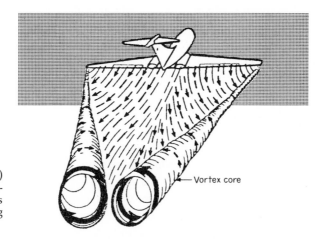

FIGURE 12-1(*a*)
The generation of wake turbulence from wing-tip vortices and downwash flow resulting from wing lift.

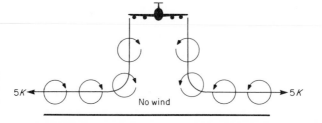

FIGURE 12-1(*b*)
Wake turbulence vortex movement in ground effect with no wind.

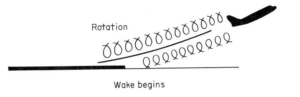

FIGURE 12-2(*a*)
Wake turbulence is associated with slow-speed, high-lift flight.

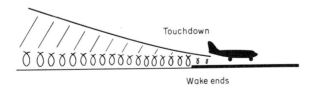

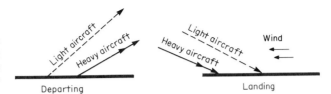

FIGURE 12-2(*b*)
Recommended departure and landing paths for small aircraft operating near possible wake turbulence.

ration, so the wing vortices are almost as large and strong as they are during an approach.

These vortices may be thought of as large swirls expanding from each wing tip, with flow rotation moving out and up from the high-pressure bottom surface near the tips toward the low-pressure area on the top surface, and then flowing aft. This motion is shown in Figure 12-1(a) and is typical of all lifting wings, although there is very little damaging energy flowing from our small airplanes. Quite frequently powerful tip vortices can be found as much as 15 mi behind large aircraft (Ref. 12.1), so do not follow too closely.

Unless a crosswind of at least 5 knots or a headwind/tailwind of at least 12 knots is blowing over the runway, disturbed flow from a landing or takeoff will tend to remain in the area for a greater period of time than the normal spacing between aircraft landing or departing at a busy airport. Recommended flight procedures to avoid these turbulent areas are given in Figure 12-2, but the addition of a few minutes time is also suggested to clear the air of stray flow patterns.

Note in 12-2(a) the position of vortex wakes behind departing and arriving aircraft. These patterns rather naturally lead to taking off and climbing out on a path staying above a large airplane until turning clear of its wake. When landing, enter and descend on a path upwind of the large airplane, thereby flying in clear air as in Figure 12-2(b). Of course, these procedures also eliminate the possibility of lingering jet blast effects which can be as serious and damaging as wake turbulence. Reference 12.2 provides more comprehensive procedural coverage for this subject if additional information is desired.

POSSIBLE CURES   Some of the airport and airways problems responsible for accidents blamed
upon pilot error would be eliminated by the following 15 recommended improvements, 13 of which could and should be put into effect as soon as
possible.

1. Airports should be free of approach and takeoff runway obstructions. This
   includes relocation or elimination of field obstructions causing runway turbulence (which endangers controllability during takeoff and landing). Although these desirable conditions cannot be met at many existing airports
   unless an interested pilot's association, enlightened FBO, or aviation-
   oriented town government takes appropriate action, future airport sites
   should be located to provide clear approaches to all runways.

   Whether large or small, airports should be surrounded by areas clear of
   housing—not because any airplane is going to drop out of the sky, but to
   provide clear approaches as well as to prevent citizen protests against airport operation. It seems to make little difference whether or not an airport
   precedes a housing development; once they settle near an airport, most
   homeowners follow the fundamental philosophy that the airport should be
   closed for their protection, and the courts seem to agree all too frequently.
   So our badly needed small airports are steadily being converted into super-
   markets and housing developments. (See Ref. 12.3 for further discussion of
   this subject.)

2. In addition to removing obstructions at runway ends, high obstructions
   around airports or along listed airways should be taken down or never installed. In fact, keeping everything over 150 ft high out of an area 3 mi
   each side of airways or a similar distance away from each side of airports
   would be ideal—probably too ideal. How many times have you encountered television or other transmitting antennas soaring up to over 1200 ft
   above ground level, or suddenly found one looming ahead during extreme
   haze conditions? Despite the most expert navigational ability, variable or
   unreported winds can drift aircraft away from airport boundaries and airways.

   There are so many collision accidents every year caused by aircraft
   striking towers, tall buildings, and other obstructions bordering airports or
   positioned along airways that legislation concerning such structures seems
   long overdue. If we can do nothing about prior installations, consideration
   should at least be given to preventing such hazardous construction in the
   future. This is another situation where the FAA and major pilots' organizations could join forces to correct a continuing source of collision accidents.

3. Small airport surface maintenance should be improved to immediately fill
   ruts, potholes, etc. that may cause taxi damage. This would be an excellent
   start toward reducing pilot error taxi accidents, with necessary repair work
   supported by local taxation and federal airport system funds. With around
   $3 billion accumulated in the Airport and Airways Trust Fund at the end of
   1978, there seems little reason to withhold assistance from any operating
   airport in need of basic maintenance work. If the government wishes to

support job training or provide broad support for unskilled and unemployed labor, repairing the runways and taxi strips of our necessary and useful small airports would be a good place to start—with funds collected from the users, the general aviation industry. As a memo to those readers flying from airports in need of repair: take a few moments to ask your congressional representatives to support your local airport; you will probably find more assistance available than you expect.

4. Windsocks and T's should be placed near the ends of runways where they can be seen and will show local threshold wind conditions (see Figure 12-3). If these indicators cannot be located at each end of every runway, every small airport should provide at least one T or sock large enough to be seen without searching the airfield. This means wind direction indicators must be of prominent color and frequently reconditioned or replaced to remain that way.

   Suitable socks can be homemade or may be purchased from many airport supply sources. Years ago, Shell, Mobil, and other oil companies offered large, yellow windsocks complete with their company name and emblem, but judging from the frayed condition of most windsocks today this practice must have been discontinued. Again, support for such airport improvement items should be provided from Airport Trust or Development Aid funds.

5. Major airports should have separate and shorter runways for small aircraft. This would help to reduce wake turbulence and jet blast problems for small aircraft and would also completely separate approach and takeoff climb speed differences so annoying to transport operation. Locating FBO facilities near these runways would eliminate confusing networks of taxiways and unnecessary taxiing over vast areas, which tends to tie up commercial transport operation while confounding private pilots. Somewhat similar treatment is now provided for helicopters, and could easily be combined with general aviation operations and facilities.

6. Runways should be numbered to eliminate or reduce frequent confusion regarding takeoff or landing instructions. As shown by Figure 12-3, it is

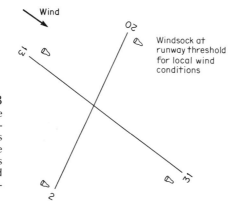

FIGURE 12-3
Confusion between "2" for the 20° runway heading and runway "20" (for 200° heading) is rivaled at this airport by the possibility of mixing directions for "13" and "31." Note wind socks at left side of all thresholds.

Wind

Windsock at runway threshold for local wind conditions

possible to unintentionally end up with a set of confusing runway headings that could have been avoided with a bit more forethought and planning. For example, a runway positioned 020 and 200 could be confused by instructions to land on "20" which means runway 200 as called.

Because of fatigue or lack of familiarity, occasional downwind landings might occur on 020, which is the 20 runway. Since runway designations are rounded within 5° to the nearest 10° heading, if this runway had been positioned along 10° and 190°, it could have been numbered 1 and 19, which is more difficult to mix up. Or located 030 and 210, depending upon prevailing wind, with subsequent landing instructions being for runway 3 or 21. Either designation is much less likely to be confused to the extent 20 and 200 could be. In similar manner it is possible to reverse the numbers 13 and 31—to take off or land downwind on the other runway. This situation showing two confusing runways on one airport actually exists in at least one location in the United States, with many other airports having one or more sets of confusing designations.

7. Unicom effectiveness would be improved for airports separated by less than 30 mi by using different frequencies in combination with reduced power transmission for all unicom frequencies. When requesting advisory information not only is it often difficult to get "a word in edgewise" with everyone talking at once, but frequently the response is also drowned out. In some areas the chatter is so bad that the FBO doesn't bother to respond to unicom communications, or, in fact, may not even have the set turned on.

By using different frequencies for nearby airports likely to experience unicom interference, we could enjoy clear reception similar to control tower operation. There should be five to ten unicom frequencies assigned with the one for each unicom airport printed on charts in the same way as control tower frequencies. To further reduce interference, when selecting unicom frequencies the transmitting power could be automatically reduced to restrict reception range. Revising unicom procedures to satisfy these requirements is not a difficult task and could be accomplished with present frequency allocations and avionics development. If phased in over a period of 5 to 10 years, by 1990 we would all enjoy unicom communication as originally intended. Present unicom is just too cluttered to be useful in today's airspace; as a result, a number of pattern, approach, and landing accidents which occur every year are blamed upon pilot error when they really originated from lack of an audible or any unicom communication.

8. Small airports noted as using unicom should monitor communications to provide a response when requested as well as to advise of local pattern activity, the active runway, and wind velocity and direction. This service could be provided by responsible volunteer help assisting on a regular schedule. After all, if yacht clubs can organize race committees recruited from active members, those of us interested in flying should be able to provide similar support.

9. Airports with control towers should always advise of any unusual or severe local wind shear, wake turbulence, or approaching storm conditions when

first contact is made. Simply providing airport barometric pressure and surface wind information is neither sufficient nor a safe practice in the presence of abnormal local conditions.

10. Communications between pilot and tower or controller should be improved through use of standard phrases more obvious in meaning than many in current use. For example, the statement "cleared for the approach" might be better stated as "cleared for final approach after crossing (the approach fix) inbound." Regardless of whether every instrument pilot is taught that "cleared for the approach" includes the responsibility to first properly execute a procedural turn and then clear the approach fix whatever it may be, during time of stress pilots infrequently flying IFR will be inclined to begin descent when cleared. (And even ATR pilots experience the same reaction, as indicated by the major airline accident near Dulles International Airport a few years ago.)

Responsibility for improving airways communications does not rest solely upon FAA personnel, but should be equitably shared by major general aviation associations, such as the Aircraft Owners and Pilots Association (AOPA), the National Business Aircraft Association (NBAA), and the General Aviation Manufacturers Association (GAMA) plus airline pilot groups working with the FAA. This is a more serious matter than it may appear to be from casual observation, with corrective action long overdue.

On the other hand, there is no excuse for any pilot under air traffic control supervision to fear "the voice on the ground." If an order is not fully understood or has come through in garbled form, do not hesitate to ask for clarification as many times as necessary to positively understand instructions. And if orders are relayed too quickly, ask for a slower repeat, remembering that controllers want to work with you, but sometimes are overly busy. However, under no circumstances should a holding point be passed without air route traffic control (ARTC) approval. If instructions are not received 5 min prior to reaching the checkpoint, request clearance, hold your position if necessary, and do not advance until so advised.

Another constant source of communications annoyance arises from automatic terminal information service (ATIS) reports that are usually so garbled it is impossible to tell what is being said without having previously read the bulletin being transmitted. Instead of helping to ease pilot workload in terminal areas, these communications often add nothing but confusion—with the tower finally being requested to supply active runway, surface wind, and barometric information, all of which were to be given by the ATIS report. Trying to understand a poor recording made by someone with poor diction leaves much to be desired. With all the good recording equipment available, there seems little excuse not to provide clear transmissions; we would all welcome any improvement making ATIS information and instructions more useful.

With general aviation IFR traffic workload increasing at a rate exceeding 43 percent between 1972 and 1976, it is essential that everything possible be done to clarify and simplify communications between air and ground. The increasing number of IFR operational accidents are silent testimony to the need to improve communications. This is a major program which should be jointly undertaken by the FAA, airline pilots' associations, and our leading private pilots' organizations.

11. The holding pattern configuration for uncongested areas could be simplified. Although I do not pretend to be an airways or air traffic control specialist, a couple of present procedures seem to be in need of simplification. The racetrack, or oval, holding pattern is probably necessary for very congested metropolitan areas, but seems complicated for most intersections throughout the United States such as Think, Conde, Monon, etc. As suggested by Figure 12-4, holding at remote points would seem to be just as effective if the intersection, omni, or beacon were merely crossed during circular flight.

    By following this procedure, small aircraft could enter a 2-min banked (IFR) turn as soon as the fix was reached. If more than one airplane was to be held at any point, 1000-ft vertical separation could be used to stack up a number of aircraft. Of course, a large number of aircraft would not normally be expected to hold in less traveled areas.

    The circular turn holding pattern would permit ease of entry, simplify wind correction by allowing the airplane to cross the fix point on every pass, and positively locate a holding airplane over the fix every 2 min. This is rarely done at most holding points, regardless of how ideal the regulations may appear. Of course, TCAs and other congested holding points need the oval pattern now required, permitting a number of aircraft to be placed at one level by extending the legs if necessary. But these areas are normally flown by pilots continually exposed to IFR operation and so more fully qualified to do the precise flying demanded by a congested holding pattern. Some study of Figure 12-4 will show the circular pattern is much simpler for the average pilot and should be adopted for remote holding points, which, after all, are the only ones most of us use. And why should right-hand turns be standard for holding when left-hand patterns are standard procedure at most airports?

12. Air traffic control should monitor flight speed of approaching aircraft and advise for time changes according to conditions. Even though most small aircraft will never have distance measuring equipment (DME) to permit

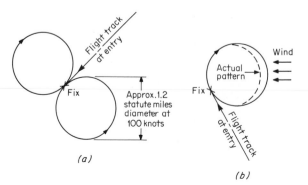

FIGURE **12-4**
Suggested 2-min-turn holding pattern for small aircraft operating in uncongested areas. (a) Whichever direction of turn is selected, cross over the fix on original omni setting and heading, left or right turns. (b) Regardless of wind effect, pattern may be easily adjusted to cross over the fix on original omni setting and heading. Right turn shown.

ground speed corrections for exact arrival times, control radar can track true speed, which may be simply computed to provide intersection arrival times. Rather than have an airplane arrive early and enter a hold, ARTC should be able to advise of necessary speed reductions to allow for wind shifts, clearance changes, etc. With this support, we would really have air traffic control.

By being instructed to adjust flight speed, we would save fuel otherwise wasted in a holding pattern. This procedure would also relieve pilots from the burden of trying to guess arrival times without knowing exact ground speed, providing assistance that will reduce both workload and opportunities for pilot error during periods of difficult IFR operation. These advantages would seem to justify the need for monitoring enroute speeds and separation, if not continually, at least during severe IFR weather in less congested areas.

13. The use of a simple HUD panel as discussed in earlier chapters would greatly simplify cross-country and IFR operation by presenting necessary flight data within the pilot's line of sight. During cruise and IFR approach, the omni track could be displayed as a series of dashed lines (see Figure 4-1), with the possibility of presenting different colored tracks for different cross-country airways within a region. Although this equipment is not presently available, such development is feasible at about the present cost of omni navigation. With some effort we could be flying similar equipment by the mid-1980s. In view of the potential market as well as the increased flight safety offered, HUD panels suitable for general aviation should be developed as soon as possible.

14. Although midair collisions are fortunately not as frequent as approach stalls and spins, there is little doubt that a basically simple proximity indicator would increase flight safety and reduce collision accidents, particularly in congested terminal areas. Discussed in Chapter 4, a proximity indicator with price and operation suitable for general aviation use should also become available during the 1980s.

15. Since weather has its own set of rules we can do little to control its activity, although we can closely monitor all movements and changes. More prompt reporting of current wind and weather information at flight service stations (FSS) would be a good start toward improving cross-country flight safety, followed by the use of area charts issued every 2 hr to cover locations of all fronts along with their direction of motion. This data would be helpful in skirting weather systems while also preventing many pilot error accidents blamed upon "inadequate preflight preparation" or "initiating flight in adverse weather conditions." When current weather type and location data are not available, any pilot is just as liable to wind up in the middle of an active front as to be successful in avoiding one. If the accident rate is to remain constant, to say nothing about decreasing, the quality of aircraft weather reporting must improve at a rate paralleling the increase in IFR operations—and that means by about 10 percent per year, compounded yearly. Few situations can overstress a pilot faster than flying into

unexpected or unpredicted severe weather, and as the cabin workload goes up, so does the accident rate.

*   *   *

This chapter has indicated operational areas that should be studied to improve flight safety. In fact, the entire field of air traffic control requires thorough review and improvement, as was highlighted by the following excerpt from the Air Line Pilots Association publication, *Air Line Pilot,* in which the members complained about "excessive communications, frequent clearance changes, lack of standardized procedures, cluttered maps and charts, inaccurate or incomplete information, complicated maneuvers, and numerous opportunities for error." And they fly the system daily!

REFERENCES   12.1   Capt. D. Leppard, Seven Miles is My Minimum, *Air Line Pilot,* March 1975.

12.2   *Airman's Information Manual,* part 1: Basic Flight Manual and ATC Procedures, Federal Aviation Administration, May 1976, pp. 1–101. Referenced data is also carried in current issues which are available from FSS and GADO (FAA General Aviation District Offices) or may be ordered through: Superintendent of Documents, U.S. Government Printing Office, Washington, D.C. 20402. (Published quarterly.)

12.3   *Potential Closure of Airports,* Federal Aviation Administration, Washington, January 1978.

12.4   FAA AC No. 150/5300-4B: *Utility Airports,* June 24, 1975. For sale by the Superintendent of Documents, U.S. Government Printing Office, Washington, D.C. 20402.

13
THE ECONOMICS OF SAFETY

In addition to the obvious benefits realized from having fewer accidents and so saving many lives, safer aircraft will also permit reductions in hull and manufacturers' product liability insurance costs, which are high now and getting higher—and are included in the price of new aircraft. Whether we realize the fact or not, we are all paying for the frequent lawsuits and settlements following many accidents caused basically by pilot error.

Recently quoted rates for owner's hull insurance on a personal landplane will be about 2.5 percent of the aircraft cost, provided the owner-pilot has established an unblemished safety record and the airplane is accepted by insurance underwriters as being relatively trouble-free. This rate seems low until you consider that coverage for a $30,000 airplane with $250 deductible damage will cost about $750 per year. If you fly 100 hr per year, hull insurance alone has added $7.50 per hour to your operating cost.

Personal liability insurance is mandatory for a responsible owner, and will cost about another $250 per year for $500,000-single-limit liability. So when flying 100 hr per year, insurance will cost $10 per hour before the propeller turns a single revolution.

To this total add aircraft manufacturers' product liability coverage, which may run as high as 3 percent of the manufacturing cost, and higher for new manufacturers or small production runs. This is another $900 of basic cost added to a $30,000 airplane—becoming $1 every hour for 9 years of flying 100 hr per year. This brings indirect operating costs up to $11 per hour without considering loan interest, and we haven't even stepped into the airplane. If the plane is flown 200 hr per year, a high time for the average pilot, it will still cost $5.50 an hour just to look at the airplane.

While aircraft insurers and underwriters tell us how cheap aircraft insurance is compared to automotive coverage premiums, this is not so if operating hours are considered. Very few people drive a car as little as 2 hr per week for 100 hr per year operation; 300 to 400 hr of driving is a more likely average totaling 12,000 to 18,000 mi per year of road travel. And there are always many cars in operation, providing constant opportunity for surface accidents.

Despite this exposure, automotive insurance premiums including collision (hull) and personal liability will be about one-fourth that for an airplane operated by the same owner. It is true that an average new automobile does not cost as much as an average new airplane, but that is not the entire reason for lower automotive rates. The cost for insuring an automobile is lower primarily because the automobile accident rate per passenger mile is less than one-tenth that of personal and corporate aircraft. According to NTSB records for 1977, general aviation deaths in the United States totaled 160 per 1 billion passenger miles vs. 13.3 for automobiles. With buses at 1.3, United States commercial airlines at 0.4, and passenger trains at 0.1 per billion passenger miles, it is not difficult to understand why personal aircraft insurance premiums are high, or why aircraft manufacturers must protect themselves with expensive product liability insurance, paid for by the eventual purchaser, of course.

Since one-tenth the accident rate tends to produce greatly reduced premiums, it seems obvious that any continued reduction in general aviation accident rates will result in a comparable reduction in premium costs—at least in constant dollar values. We must act to bring about this improvement as soon as possible if flying costs are to be kept within reasonable levels for the average pilot.

In this regard, George W. Meehan, senior vice president of United States Aviation Underwriters, one of our oldest and largest aviation insurance companies, recently commented that he "would expect, in the near future, that insurance rates on privately owned aircraft will escalate inasmuch as 1977 and the first 6 months of 1978 have shown a marked increase in general aviation accidents."

He continues: "We are in the age of consumerism and the manufacturers' product liability exposure will continue to grow in direct proportion to the ever-increasing sale of aircraft. Increased claims activity will cause product liability insurance costs for aircraft and component manufacturers to dramatically increase." So much for the future, if the accident curve cannot be turned down.

This pessimistic projection can become optimistic, however, if steps are taken to make flying safer. While it is impossible to legislate or mandate common sense, it is possible to preclude pilot error by "designing out" features or procedures that cause or permit pilot error accidents. This book includes many safety recommendations that could be incorporated into existing and new aircraft to considerably reduce accident frequency.

LITIGATION COSTS We can also reduce purchase and flight costs by taking positive action to discourage the number of irresponsible lawsuits being brought against manufacturers of aircraft and components. Before a liability claim can be imposed, it should be necessary to prove that: (1) the product in question is or was defective; (2) the defect existed when the product left the manufacturer; (3) because of the defect the product was unreasonably dangerous to the user of the aircraft; (4) the user of the aircraft was injured or suffered damages; and (5) the defect (if proved) was the *real cause* of the injuries or damages proved to have been suffered. Many published settlements today seem to neglect or purposely overlook these basic requirements for a judgment, resulting in increased airplane and flight insurance costs (Ref. 13.1).

There are two further points that should be noted in relation to any legal action arising from an aircraft accident. Referring back to English law, we should give consideration to the proposition that no injury is done to one who consents. In other words, pilots and their passengers should possess common knowledge that flying is potentially more dangerous than driving a car, and thus they are exposing themselves to a greater degree whenever they fly. From this viewpoint, aircraft accidents should not be totally unexpected, and this knowledge should considerably dull any sharp-edged claim.

For the second point, a recent court decision emphasized what should be obvious, but apparently is not to many: "A consumer would not expect a Model T to have the safety features which are incorporated in automobiles made today. The same expectation applies to airplanes. Plaintiffs have not shown that the ordinary consumer would expect a plane made in 1952 to have the safety features of one made in 1970." This means that injuries or damages experienced by occupants of an airplane certified in 1952 are not grounds to sue because the airplane was not equipped to 1970 FAA standards which might have prevented such injuries or damages (Ref. 13.2). By the same reasoning, it should not be possible to bring to trial any lawsuit claiming that injuries would have been prevented if "such and such" unrequired equipment had been installed.

If equipment can be shown to save lives in accidents, it should be used, but it is not mandatory unless required by federal regulation. (The recent FAA amendment installing shoulder harnesses is a case of mandated improvement, although for maximum safety this regulation should be extended to cover all passengers in addition to just those in the forward seats.)

Adherence to these fundamentals should prevent irresponsible product liability awards, each of which costs every pilot just a little bit extra to fly; unfortunately, we have recently seen many such judgments that collectively cost us more and more to own and operate an airplane (and our automobiles as well).

In conclusion, a constantly decreasing accident rate should keep aircraft purchase and operating costs near present levels. Thanks to dollar devaluation this implies an actual reduction in insurance rates and aircraft list prices, providing lower operating costs. As a result, not only will more owners be able to afford new aircraft and related equipment, but prices of secondhand aircraft will stabilize and eventually be lower. This in turn will expand the aircraft market by making ownership appealing to pilots who could not previously afford high purchase, maintenance, and insurance costs that became unjustifiable operating costs.

$$*\qquad*\qquad*$$

As a postscript, it is my sincere hope that pilots, owners, manufacturers, and supervisory FAA personnel will eventually realize the growth possible for our industry through improved flight safety and personal aircraft utility, and then work together to attain the level of acceptance small aircraft deserve but have never enjoyed. If this book contributes to that goal, it will have served its purpose. Possibly you, too, can take some action to make these safety features become reality and so contribute to more relaxed and enjoyable flying. Continued growth of general aviation is worth the effort required to provide the next generation of safer and more efficient aircraft and operational procedures; indeed, tomorrow would not be too soon to begin this new chapter. By eliminating many opportunities for pilot error through design and operational procedures discussed in this book, we can take a major first step in that direction.

REFERENCES   13.1   548 Fed. 2nd 288, 1977.

13.2   *Bruce v. Martin and Ozark.* United States Court of Appeals, Tenth Circuit. September 24, 1976.

APPENDIX

This appendix has been included to provide background information of general interest concerning the major aviation safety organizations as well as a review of current procedures for accident investigation.

**1
SAFETY AGENCIES**

The primary overseer of aviation safety is the National Transportation Safety Board (NTSB), an independent federal agency entrusted with the mission of safety improvement for all types of transportation including railroad, highway, pipeline, marine, and civil aviation. Originally organized under the Department of Transportation, the NTSB became totally autonomous on April 1, 1975, at which time its investigative powers were also increased in scope.

A brief comment about the relationship of the Federal Aviation Administration (FAA) and the Civil Aeronautics Board (CAB) to NTSB investigation, analysis, and reporting of civil aviation accidents may be of interest, particularly since the CAB used to direct aircraft accident investigations as well as issue related findings and statistical reports. According to Edward E. Slattery, Jr., director of the NTSB Office of Public Affairs who kindly supplied much detailed information about NTSB operations and procedures, FAA accident responsibility falls into two categories: (1) assisting the NTSB with accident investigation work, and (2) enforcing any violations of the United States Federal Air Regulations (FARs).

The CAB is no longer involved in aircraft safety or accident investigation, having held this responsibility from 1940 to 1967 when the Safety Board (NTSB) was created as an arm of the Department of Transportation. At the present time, the CAB is concerned with air transport routes and rates plus the economic well-being of the passenger/cargo operations of certificated route and supplemental air carriers, and so does not participate in aircraft accident studies.

Since we are primarily concerned with the general aviation segment of NTSB operation, it is interesting to note that the five-member Safety Board plus related support personnel are now empowered to: investigate and determine the probable cause of aircraft accidents, review and analyze accident data, evaluate the performance of other government agencies such as the FAA, conduct special safety studies and investigations, publish accident reports and approved safety procedures, and submit proposed legislative changes as well as safety recommendations to Congress. These broad powers provide a base for exerting considerable influence upon general aviation design and flight operations, although the FAA frequently seems to respond extremely slowly to NTSB safety policies or requests for specific changes. As a result, a safety finding may be years in implementation with many lives needlessly lost before a desirable change is effected, viz., the Turkish Airlines DC-10 cargo door accident of March 1974 which occurred more than a year

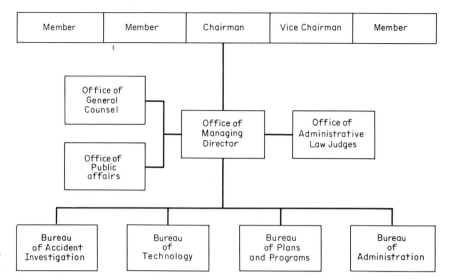

| Member | Member | Chairman | Vice Chairman | Member |

Office of
General
Counsel

Office of
Public
affairs

Office of
Managing
Director

Office of
Administrative
Law Judges

Bureau
of Accident
Investigation

Bureau
of
Technology

Bureau
of Plans
and Programs

Bureau
of
Administration

FIGURE A-1
National Transportation Safety
Board organizational structure.

and a half after the NTSB requested the FAA to mandate immediate structural design modifications in the interest of flight safety (Ref. A.1). I believe the moral of this example is simply that adding another layer of bureaucracy has never resulted in an increased rate of action. However, this accident did prompt the independence of the NTSB from the Department of Transportation during the following year.

The accident investigation procedures and reports published by the NTSB will be briefly reviewed further along in this appendix.

The FAA also contributes most positively to general aviation safety. In addition to the mandatory 100-hr (for commercial operation) or the annual aircraft inspection requirements, which frequently reveal potential trouble before a failure occurs in flight, the FAA employs field personnel who are a ready source of maintenance tips, accepted repair procedures, and suggestions for safe operational practice. Through the *Inspection Aids* bulletin issued monthly by FAA personnel, recent structural or mechanical failures are described for various aircraft models along with photographs or sketches of the defective parts, hours of operation at which the failure occurred, and recommended corrective action to be taken.

The FAA is authorized to issue Airworthiness Directives (ADs) whenever a defect is considered to jeopardize flight safety. Written to cover a specific problem experienced with a certain aircraft type, the AD carries sufficient legal weight to ground an airplane if necessary until the required modification has been incorporated. Needless to say, aircraft manufacturers view ADs as a mark against their design or manufacturing procedures, and take pride in having no or very few ADs issued against their products. Of course, product liability insurance, lawsuits, and hull insurance also increase with the number of ADs against any design. During the course of this book, we have reviewed design procedures intended to ensure long aircraft life while minimizing AD and accident potential. Unfortunately, service use and maintenance (or lack thereof) frequently impose conditions unanticipated during design and manufacture, and so an AD becomes necessary.

To digress a moment, it is important to note that an AD may not always indicate a design or manufacturing deficiency; it may be necessary to reinforce structure or revise manufacturing procedures because of operating life, service use, or environmental conditions not intended or anticipated by the original aircraft design specification. The required Beech 18 wing spar reinforcement is a problem of this type, as are modifications of landing gear support structures frequently implemented on aircraft originally intended for runway operation but subsequently found to perform well in rough (and rougher) bush-country terrain. As a result, FAA enforcement of safety through the use of ADs will remain with us as long as aircraft are built.

Of course, the high level of design strength, equipment quality, and flight handling characteristics required by FAA Design Type Certificate (TC) specifications represents a basic contribution to flight safety, as discussed in *Design for Flying* (Ref. A.2).

In the private sector, the oldest and most widely recognized aviation safety organization is Flight Safety Foundation, Inc. First formed in 1945, but really activated by aviation insurance engineer Jerome Lederer in June 1947, Flight Safety Foundation is a nonprofit organization headquartered in Arlington, Virginia. Over the years the activities of this group have grown to include formal aircraft accident investigation instruction, the development of mechanical malfunction reports now issued by the FAA, recommendation of the use of in-flight recorders and anticollision lights, the first crash fire rescue programs, standardized pilot training, and safety research studies.

Affiliated with the Safety Center of the University of Southern California, Flight Safety Foundation is supported by approximately 450 private and corporate members, including 93 of the world's airlines. Additional income is realized from staff publication of safety bulletins, newsletters, and books, as well as reports of the two major seminars held each year on special subjects related to international air safety and corporate aircraft operational safety.

Acting as participant or sponsor of eight annual awards for significant achievements in aviation safety, Flight Safety Foundation has been similarly well recognized over the years by receiving over 40 awards for its own contributions to aviation. Called the "conscience of the industry" for quietly correcting aviation safety imperfections and uncertainties with practical solutions, Flight Safety Foundation provides a constant and steady force working to improve safety at all levels of civil aviation, and in the process exercises considerable influence internationally as well as here at home.

Leader among the pilot organization safety groups is the Aircraft Owners and Pilots Association (AOPA) Air Safety Foundation. Supported by members' contributions, the Air Safety Foundation publishes the quarterly *Flight Instructors' Safety Report* carrying recommendations for improved flight training and operational procedures; prepares private aircraft accident reports and analyzes data for publication in *The AOPA Pilot*; sponsors private pilot and instructor refresher courses on a regular schedule across the country; prepares and distributes films concerned with flight safety practices; acts as an independent source for compilation of general aviation operational, accident, and pilot rating statistics; and conducts safety research studies in the field of general aviation. In fact, during 1977 the AOPA Air Safety Foundation received the FAA's Extraordinary Service Award for advancing aviation safety through pilot education, aviation research, and general aviation safety standards projects.

With over 45,000 members, many of whom are enthusiastically building their own aircraft from one of a number of different designs and construction kits available, the Experimental Aircraft Association (EAA) has recently formed an Office of Aviation Safety. This branch of the organization provides guidance for pilots planning that "first ever" flight of their newly built airplane, studies and recommends proper procedures for safer aerobatic flight in the many homebuilt aircraft constructed for that purpose, acts as a central clearinghouse to maintain records of active EAA homebuilt aircraft, and analyzes and distributes data related to amateur (homebuilt) aircraft accidents. With estimates of up to 10,000 such aircraft under construction at any one time, the EAA Office of Aviation Safety serves a valuable communications function highlighted by special forums offered each year for various models of homebuilt aircraft. Held in Oshkosh, Wisconsin, starting on the last Saturday in July with over 300,000 aviation enthusiasts in attendance over an 8-day period, the EAA Annual Convention provides a unique platform for distributing safety recommendations and concepts to the EAA membership. A similar, though smaller, EEA convention is also held annually at Lakeland, Florida, during mid-March and includes useful forum programs.

In addition to these principal safety organizations, there are many other companies and consultants specializing in aviation safety via research studies, product development, or accident investigation. With all this interest focused upon aviation safety, it would seem logical to expect our accident record to improve most rapidly. But it does not, primarily because we are studying accident causes and effects when we should be focusing our attention upon ways to eliminate or at least reduce causes of pilot error accidents over which we have some degree of control. So, considering current statistics plus the fact that accident investigations will always be necessary thanks to unpredictable weather—even if we could eventually eliminate all possibility of pilot error—let us take a look at how aircraft accidents are studied in the field.

2 ACCIDENT INVESTIGATION

While the NTSB is directed by law to determine the cause of all air transport accidents in the United States, a function that cannot be delegated to any other agency, the investigation of nonfatal lightplane and helicopter accidents may be conducted by the FAA with authorization and approval of the NTSB. Even with this delegation, the NTSB retains the statutory duty to determine probable cause of the accident in each case. Foreign accidents involving aircraft of United States registry or manufacture may be investigated by an accredited representative of the NTSB in accordance with recommended practices of the International Civil Aviation Organization (ICAO).

In order to reach a fatal air transport crash site as soon as possible, the NTSB maintains a Go-Team in constant readiness at its Washington offices. Composed of as many as 10 investigators with specialized aircraft background and accident training, Go-Team personnel are rotated weekly and remain on call day and night when on duty. During the investigation of any major United States air disaster, the Go-Team is usually accompanied by at least one NTSB board member, and may work from 5 to 10 days at the accident site gathering factual evidence for presentation at the public hearing held for each major air transport accident. Such investigations will normally be assisted by FAA specialists trained in accident analysis.

There are 12 NTSB field offices located across the country, one of which is usually requested to handle general aviation fatal accidents occurring in its area. One investigator is in charge of each fatal general aviation case, with specialists in any category of skills supplied from Washington headquarters if requested by the field office. Regardless of any assistance supplied by the FAA during investigation of nonfatal or fatal general aviation accidents, the probable cause of such accidents is usually decided by the NTSB without a public hearing and as based upon analysis of the chief investigator's findings. If pilot discipline is required in connection with general aviation accidents, the FAA will normally initiate proceedings after determination of the probable cause of the accident.

These legal proceedings in fact are based upon several long and tedious congressional acts which form the Federal Aviation Act of 1958, as amended, and the more recent "Government in the Sunshine Act," as amended, which is primarily concerned with access to public records, or more correctly, public hearings, for our particular field of interest. While I do not wish to dig into the nitty-gritty of ambiguous public law jargon, a few of the legal requirements of accident reporting and investigation will be of interest to those curious about this phase of aviation.

For example, consider a portion of NTSB Regulations Part 830, Rules Pertaining to the Notification and Reporting of Aircraft Accidents or Incidents. . . . You might be surprised to learn the following, in paragraph 830.5:

The operator of an aircraft shall immediately, and by the most expeditious means available, notify the nearest National Transportation Safety Board field office when:

(a) an aircraft accident or any of the following listed incidents occur:
    (1) Flight control system malfunction or failure;
    (2) Inability of any required flight crewmember to perform his normal flight duties as a result of injury or illness;
    (3) Turbine engine rotor failures excluding compressor blades and turbine buckets;
    (4) In-flight fire; or
    (5) Aircraft collide in flight.
(b) An aircraft is overdue and is believed to have been involved in an accident.

Note that these requirements were written to cover all civil aircraft of the United States. More to the point, by law we are apparently supposed to report general aviation accidents to the nearest NTSB field office and not to the FAA. Just in case you ever need to file an accident report, the NTSB field offices are listed under United States Government in the telephone directories of: Anchorage; Atlanta; Chicago; Denver; Fort Worth; Kansas City, Missouri; Los Angeles; Miami, Florida; New York; Oakland; Seattle; and Washington, D.C.

NTSB Part 831, Rules of Practice in Aircraft Accident/Incident Investigations, notes in paragraph 831.2 that the "Board is solely responsible for the investigation of all accidents involving civil aircraft, or civil and military aircraft, within the United States, its territories and possessions." This paragraph also states that the board is responsible for investigating accidents involving United States civil aircraft, or civil and military aircraft outside the country when so permitted by international policy and agreement. The fact that the NTSB has the power to investigate accidents between small and military aircraft should be encouraging news to any of us who have unwillingly served as a target run for playful military pilots.

Another interesting point, raised by paragraph 831.5, is that "No party to the field investigation . . . nor any party to the hearing phase of an accident investigation . . . shall be represented by any person who also represents claimants or insurers." Simply stated, this means that anyone who has a claim against parties involved in an accident may not investigate or testify (for personal gain)—certainly a desirable approach to help prevent self-protecting and improper accident investigations.

Once on the scene the investigator-in-charge is authorized to organize, conduct, and control the on-site investigation. This includes responsibility for recruiting parties to the field investigation, plus supervision and coordination of all resources as well as the activities of all NTSB and other personnel involved in on-site work.

With broad legal authority, authorized on-site representatives of the NTSB may question any person having knowledge relating to an aircraft accident or incident. They are also the only people legally provided access to the aircraft wreckage, mail, and cargo. Release of information during a field investigation can only be made through the board member present at the accident scene, the representative of the board's Office of Public Affairs, or the investigator-in-charge.

Following the on-site study and gathering of related data covering the aircraft, flight personnel, weather conditions, wreckage resulting from the accident, etc., the NTSB may schedule a public hearing to create a public record of the facts, conditions, and circumstances relating to the accident. Such hearings are purely fact-finding proceedings, without formal pleadings or issues and with no adverse parties. As such, these hearings constitute a formal board of inquiry under the direction of an NTSB board member who chairs the proceedings, or under another member of the inquiry board so designated when the NTSB member cannot be present. As a matter of interest, any person may submit recommendations for proper conclusions to be drawn from the testimony and exhibits submitted at these hearings.

Verbatim transcripts of all hearings are available from the NTSB Office of Public Affairs or from the court reporting firm preparing the record. All this information is finally digested in a detailed NTSB report setting forth the probable cause of the accident along with any appropriate recommendations developed during the investigation.

Once an accident report has been issued, only written requests for reconsideration or modification will be accepted by the NTSB, and then only if new evidence can be presented or if the board's findings can be shown to be erroneous. This material is continuously gathered together in a file known as a *public docket,* which is prepared for each accident investigated and may be reviewed by any interested person at the Washington, D.C., offices of the NTSB. In fact, upon payment of the cost of reproduction, copies of docket material may be obtained from the Public Inquiries Section, Bureau of Administration, National Transportation Safety Board, Washington, D.C. 20594.

All this lengthy legal procedure may consume much time before an accident file is completed; but according to regulation 831.41, "Accident investigations are never officially closed but are kept open for the submission of new and pertinent evidence by any interested person." But what we *really* need is fewer accidents.

REFERENCES   A.1   Paul Eddy, Elaine Potter, and Bruce Page, *Destination Disaster*. Quadrangle, The New York Times Book Company, New York, 1976, p. 152.

A.2   David B. Thurston, *Design for Flying*. McGraw-Hill Book Company, New York, 1978, chap. 17.

GLOSSARY OF ABBREVIATIONS AND ACRONYMS

A&P	airframe and powerplant (mechanic)
a.c.	aerodynamic center
AC	Advisory Circular (issued by the FAA to inform the aviation public of non-regulatory material of interest)
AD	Airworthiness Directive (issued by the FAA to mandate aircraft inspection, repair, or modification based upon operational problems)
ADF	automatic direction finder
AGL	above ground level
Ag plane	agricultural airplane
AOA	angle of attack
AOPA	Aircraft Owners and Pilots Association
ARTC	air route traffic control
ATC	air traffic control
ATIS	automatic terminal information service
ATR	airline transport rated
CAA	Civil Aviation Authority (British)
CAB	Civil Aeronautics Board
CAM 3	Civil Aeronautics Manual 3
CAT	clear-air turbulence
c.g.	center of gravity
CID	carburetor ice detector
Denalt	*Den*sity *Alt*itude computer
DG	directional gyro
DME	distance measuring equipment
EAA	Experimental Aircraft Association
EGT	exhaust gas temperature
ETA	estimated time of arrival

FAA	Federal Aviation Administration
FAR	Federal Air Regulations
FBO	fixed base operator
FSS	FAA flight service station
GAMA	General Aviation Manufacturers Association
g loading	total weight on wings divided by airplane weight
HSI	horizontal situation indicator
HUD	head-up display
IA	inspection authorized (mechanic)
ICAO	International Civil Aviation Organization
IFR	instrument flight rules
mag	magneto (on engine)
mph	miles per hour
NACA	National Advisory Committee for Aeronautics
NASA	National Aeronautics and Space Administration
NBAA	National Business Aircraft Association
NTSB	National Transportation Safety Board
OAT	outside air temperature
omni	omnidirectional radio range or omnirange station
prop	propeller
psi	pounds per square inch
ROC	rate of climb
rpm	revolutions per minute
STOL	short takeoff and landing
TC	FAA Design Type Certificate
TCA	terminal control area
TIA	Type Inspection Authorization
unicom	radio communication at uncontrolled airports providing airport advisory service
VFR	visual flight rules
VOR	very high frequency omnidirectional range station
VTOL	vertical takeoff and landing

INDEX